Thomas "TJ" Loftin

7/19

Monetizing Gentrification

PREFACE

LEVERAGING YOUR POWER!

The name of this book is called "Monetizing Gentrification." This particular first edition is about America, all of the things that are going on in and around America. The good and not so good ways to make money.

Monetizing Gentrification to me suggests the different ways to make money during gentrification. We've got a major disruption with business and real estate changing the way businesses run and who owns real estate. This affects everyday people; people doing well and people not doing so well. So, let's talk about some of these decisions that are affecting everyone, everywhere.

ISBN: 978-0-578-20933-3

DEDICATION

I dedicate this book to my father, Mr. Thomas Loftin, (RIP). He was one of the best diesel mechanics on the planet and to my grandfather, Mr. Darvin Watkins, (RIP). He gave me my first red toolbox at the age of eight years old. He showed me how to do everything from changing the seal at the base of the toilet to making door keys. I'm thankful they showed me the power behind just owning a toolbox with tools in it and knowing how to apply it. I truly thank them for passing down the generational wealth of knowledge that I was able to use as a pathway to wealth and success through my years as a successful auto industry guru, and created the man that I am today. This knowledge has also opened the door for me to pass it forward to the next generations with my workshop called TJ's Toolbox, in order to keep the legacy going!

Chapters

FORK-IN-THE-ROAD QUESTIONS

So, we have a person standing here with his hand on his head. He's scratching his head looking at this fork in the road trying to figure out so many decisions to make:

- Should I put my children in college seeing that my other child has already graduated and can't find work?
- There are over 95% of college students currently unemployed. So he is like, Should I do that, too?
- Should I sell my business, my building that I'm currently in, or my apartment building that I own? If I do, where should I go?
- I can't really afford to buy anything, not in this economy. Should I buy a home or should I wait?
- What I can afford to buy right now, I really don't like. I don't like the area. But if I wait until next year, then I should be able to buy something much better. But I heard "TJ" say that I shouldn't wait and that I should buy whatever I can afford now while I can.
- Should I start a business? I heard a lot of bad things about people who started their own businesses and they failed. It's not really common in our community to see black people with businesses, they say. So a lot of people don't know a lot of black business owners. They always question should they start a business and would it be successful.
- Then you have people saying, "It's too expensive in my state. Should I move out of state?" We have a lot of situations going on in our families, around the world really, especially in America. So let's move forward and see what's going on.

MY HUMBLE BEGINNINGS

Let me tell you a little bit about me and why I am able to write this book and tour the country speaking about gentrification and how to monetize it.

I started off as a car builder from Compton, California, building cars and selling them. Flipping cars in my mom's backyard, building low riders, custom cars, classic cars, and regular cars anything I can make good money off of.

So, I was the little neighborhood terrorist guy that you probably didn't like because I was blasting my radio; but I grew up in Compton. I knew a lot of people and a lot of my friends became popular and became very successful. For example, when NWA (famous rap group) got into the rap industry, they started going back into the community and bringing the low rider guys in.

"Low riders" were the biggest thing going on in Compton, California, such as customs cars, classic cars, etc. So when NWA became popular, they put those cars in their videos and those videos went viral. It opened up a lot of doors around the world. So what kind of door did it open up? It opened up the door for people like me who built cars in their mom's backyard. A lot of people started off in their mom's backyard and their parents' garages. ***Express Gold Plating*** originally started off in my mom's backyard.

A lot of other famous people started off in their parents' backyards, garages, and kitchens as well: Apple, Google, Amazon, FUBU, Harley Davidson, Walt Disney, Mattel, etc. So I mentioned all of them to say this, when you hear of a young person starting a business from home, now you have some references to say "Oh yeah, I know a lot of successful people that started from home that did not have any money. They were struggling." Apple could not raise any money to save their lives to make this dumb idea called a "computer," a "personal computer." Who would possibly want one of those? Go figure.

So **"TJ"**, the brand, became successful and I started building a lot of cars because of all of those videos that were being created.

I was one of the few African American car builders in America, in the world! I became one of the best, so when I started my shop, ***"Express Gold Plating"*** in Gardena, California, in 1991, I did very well financially. I went back to the industry and I started dotting those "i's" and crossing the "t's" and I realized that these low rider cars had wire wheels. Every Sunday we would drive our cars. Most likely somebody is going to bump their car's wheels up against the curve or in a pothole. They are going to be jumping the car, their wheels are going to come down crooked and dent the wheel or mess it up and those wheels are very expensive. So at a cost of $1,000 per wheel (back then), I decided that maybe I should try to figure out a way to fix these wheels. They were so expensive and I had a lot of people coming to me every Monday asking me about their wheels.

So after I figured out how to fix wheels, **Street Customs Magazine** wrote an article about me. Unfortunately, they cut my body out the picture because they didn't want people to know an African American figured out how to fix these wheels because I was shipping wheels all around the world.

Now because I learned how to take those wheels apart, I started updating those wheels and putting colors in the wheels: green, red, blue, yellow, etc. That opened the doors for a huge industry. At that same time I also ended up getting all of the parts to make brand new wheels. That's when I decided to start my own wheel company called *"Compton Wire Wheels."* I would eventually ship wheels all around the world!

QUESTIONS

- Do you have a business idea that you would like to start working on now?

- TJ started a business without any money; do you know of any other businesses that you can start without having any money?

My relationship with shipping cars internationally started with Japan because they watched those rap videos and they wanted to be a part of the culture. I shipped everything from cars to clothes because they were intrigued with our way of life in Compton. I sold cars to very successful people whose children liked low riders, and their grandparents owned a lot of successful businesses. You would be surprised who is driving low riders in Japan.

These are some of my oldest Japanese customers here. They have shipped thousands of cars to Japan themselves. This picture was taken at the funeral of the owner of *Heavy Weight Records* and celebrity car builder, Mr. Terry Carter. He was one of my oldest friends and one of my business coaches. (Mr. Carter was the gentleman that Suge Knight accidentally ran over on the set of the hit movie, Straight Outta Compton.)

A lot of people don't know that a low rider Japanese magazine exists, but over there they call me "Mr. TJ." I have shipped a lot of cars overseas and most importantly a lot of car parts and accessories. Many people don't realize that a lot of foreigners have low riders. There are many other countries such as New Zealand, Australia, Norway, France, Germany, etc., that purchase low riders. I became really famous for shipping lots of parts all around the world. So, because of all my history in low riding, I decided to start teaching it.

QUESTIONS

- TJ and his friends made lots of money selling and shipping low riders and low rider parts all around the world including countries that didn't have low riders. Have you ever traveled to other states or countries that didn't have something that another place did have?

- Do you know that by simply bringing a product from one place to another can create hundreds if not thousands of jobs?

- Give some examples of different types of jobs that can be created that can help society that no one has ever seen before.

I have been in countless magazines. I have also been on the cover of magazines. Due to the extra exposure around the world, I opened up my second location. I had two (2) stores: one in Gardena, California, and one in Palmdale. This unlocked the door for me to start opening more stores and I did very well.

Because of these businesses, I was able to go back to my community and hire a lot of my friends and help them start their own businesses. Due to supply and demand, I was buying so many cars in bulk. I would put my friends in tow-trucks or car carriers and send them all over the country to pick up the junk cars and bring them back to me in California.

Back then when I owned two shops, I was the biggest African American low rider shop in the world! Because of my success, I had a lot of people calling this the new craze. Low riders were starting to boom. So they decided that they wanted to start putting us in movies, videos, TV shows, commercials, and magazines. Even books were written about us.

We did this movie here called "Driving While Black."

We also placed low riders in the movie called "Straight Outta Compton." There were over 200 low riders in that movie.

I decided to purchase a house up in the Antelope Valley on acreage. That was a place where I could put all of my tow trucks, car carriers, boats, and everything in one place.

I was able to consolidate all of my cars into one location and that saved me thousands of dollars a month in storage fees.

This is the backyard of the house where I would build the supercars, the celebrities' cars.

When I talk to people about the business of "the auto industry," that refers to all genres of people in general. It's very important that people realize it's not about what's going on in your community or on your block, this is about business as a global industry. I remember before I started advertising, I was making about $5,000.00 a month and I thought I was doing very well. But my buddy said, "You have people coming from all around California that's driving 2 to 3 hours away to buy stuff from you; you should run an Ad." I thought to myself, I already make $5,000.00 a month, why would I do that? That doesn't make sense, they already know who I am. He said, "You should run an ad." So, I ran an ad and paid $500.00 a month that I did not want to spend, but when I paid that $500.00 it was for a local ad. All of a sudden I started making $15,000.00 a month immediately and that blew me out of the water!

I found out about all of these different ways of advertising. I advertised in a magazine that was sold in 4 different countries around the world and had over 1.2 million subscribers. So when I did that, it really opened the door for me to really do very well. I just became an international business person. I made more money overseas than I made in person.

People would call me from all around America saying, "Hey I have a bunch of those junk cars on my land that you are building for all of those commercials and videos, movies, TV shows, etc. Are you interested in purchasing?" and I said, "Of course!" I would pay $300-$500 for each car. I bought this three-

car carrier and had my buddies from Compton pick up those cars. I would give all of those guys jobs and help them start their businesses. Eventually, the business of picking up those cars and delivering those cars got so big it was out of control. I had so many trucks that I had to release them to my friends and let them take control of that stuff because it just got so big.

I would buy them in Mississippi, Georgia, North Dakota, South Carolina, you name it and I would send a truck. All of a sudden when they got back to Compton they would be worth $10,000.00 apiece "as is" because out of state there was no market for those cars that were not finished but in Compton, California, there was a huge market for those cars completed. Some people I know today are still buying car haulers, and they are shipping cars and are still making a lot of money finding classic cars.

So this is one of my ex-employees and this was actually the first person that

ever called me and told me, "Thank you. I've reached a level of success that I didn't think was possible. Thank you for inspiring me to start my own business.

Anyway, after all that time, I decided I needed to start giving back. So, I started teaching.

I went to *Leimert Park, California, where I heard that a lot of African American people hung out. I made DVDs so I could sell a product at the park. So I noticed the young people hanging around the park with nothing to do while their parents were vendors and sold things. I reached out to all of those young people and said, "Come on, sit down, and let's talk business. Let's go sit on the grass under the learning tree." So when I started doing that, all of a sudden I had a doctor friend of mine and other people start coming out because they were interested. I was not only talking about the car industry, I was also talking about the movie industry and TV industry. I was talking about starting businesses and how to make money. That opened up a lot of doors and a lot of people heard about it. That's when I eventually found out that they were trying to gentrify Leimert Park, California. So, I decided to organize the vendors to come together and make sure everybody was selling stuff at the same price so no one was undercutting anyone else. I just organized the whole park and kept them from being gentrified.

*Leimert Park is a residential neighborhood in the South Los Angeles region of Los Angeles, California. It was developed in the 1920s as a master-planned community featuring Spanish Colonial Revival homes and tree-lined streets. *Wikipedia*

Calling Out Gentrification in Leimert Park

After realizing people were trying to kick the vendors out of Leimert Park, I found out that the people behind it had meetings every Monday morning at 10:00 am – 12:00 pm. So people with jobs could not attend and did not know what was going on in their area. Therefore, I started attending the weekly meetings at The Vision Theatre in Los Angeles previously owned by Marla Gibbs (actress) from the hit TV show "*The Jeffersons*." As a man of the trades and a seasoned business owner, I instantly saw the deception in the room and recognized it as gentrification. So I was the first to call it out and accused the chair members of being a part of it.

I also called out Metro as the biggest cause of it by building a train under Crenshaw Blvd., which at the time most people did not know that this was happening. After two meetings of debunking all of the false stories being told by the chair members, they decided to give me my own meetings on Wednesdays at The Vision Theatre at 4pm in an effort to keep me out of their meetings. But I still went to both. If I had a regular job, none of this would have been possible.

Founder of the Leimert Park Vendors Association -'The Vision Theatre'

QUESTIONS

- Why do you think it was important for TJ to help the small business owners (vendors) to not be kicked out of the park?

- Do you think it's possible for an eight year old selling candy to his friends at school to grow up and become a billion dollar candy company owner with all of his friends working for him?

*Al Jazeera Media Network is a Middle Eastern multinational multimedia conglomerate, and is the parent company of Al Jazeera and its related networks. *Wikipedia*

After organizing the vendors and getting to the table with city members, I quickly realized this was not just about Leimert Park, this was about Black

America. So, I reached out to Black businesses and home owners in the area and explained what was going on and they started attending my meetings. Back then most people did not know what gentrification was and no one was giving meetings or talking about it to the communities in LA County. Because of the rapid growth of the meetings I was giving, Al Jazeera came out and did an interview on me in the park (Leimert Park) across the street from The Vision Theatre in Los Angeles.

QUESTIONS

- Why do you think the chair members got angry with TJ about calling out gentrification in Leimert Park?

- Why do you think investors want to keep gentrification a big secret?

I became a celebrity business coach. I met a lot of great people including the world's biggest doctor on the planet that cured all diseases, *Dr. Sebi. He came to me and said, "Hey, they don't like you, they don't like me, let's be friends." I went on to become his business coach.

*Dr. Sebi – Pathologist, Herbalist, Naturalist, and Healer *Wikipedia*

QUESTIONS

- What did Dr. Sebi become famous for helping people around the world?

- Why do you think Dr. Sebi came to Leimert Park to support TJ after calling out gentrification?

I had an opportunity to coach and consult numerous people over the years.

Isaiah Cooper, the first African American to fly around America by himself at 16 years old. He went on to become one of America's most influential African Americans. I've been coaching Isaiah before he made his amazing flight.

QUESTIONS

- How old was Isaiah when he made his historical flight around America?

- Why do you think it's important for us as African Americans, or Black people, to document our own history?

*Dr. Umar Johnson and I were in a private meeting at Leimert Park.

*Dr. Umar Johnson is a Doctor of Clinical Psychology and Certified School Psychologist and well known global speaker. After our private meeting held, he mentioned that he'd never been to Leimart Park. So we walked down the street and continued to talk on the historic fountain. *Wikipedia*

QUESTIONS

- Around how many low riders were in Compton, LA, and Watts when TJ first got started?

- What are some of the countries TJ sold and shipped low rider car parts to?

- Why did some magazines remove TJ's face from the picture in the articles written about him?

- What was the name of TJ's wheel company that he started?

- What was the name of TJ's first shop; the business he opened?

- What other famous businesses started from inside of their home?

- What city did TJ grow up in?

- What year, make, and model car was TJ sitting on in the beginning of his book?

- What type of custom work was TJ doing with cars?

- What made TJ write this book?

- What foreign country wrote a two-page article on TJ?

- What movies had low riders in them?

- Why did TJ buy a house in the Antelope Valley and why was that very important?

- Why did TJ buy trucks and car carriers?

- Who did TJ hire to drive his tow trucks?

- What did TJ teach the children while he was in Leimert Park?

- What foreign news company interviewed TJ for helping the vendors?

- What famous doctor did TJ help with his business?

- What was the main lesson you learned from reading about TJ?

REAL ESTATE TERMINOLOGY YOU NEED TO KNOW:

Adjustable rate mortgage (ARMs) – This type of mortgage usually has a lower initial rate (for a set number of years), then the rate may go up or down, depending on the specified index rate used for determination. Usually preferred for short-term ownership, the repayment period for ARMs are typically 5, 7, or 10 years, but they can be issued for longer time periods.

Amortization – The repayment schedule of a loan, including payments of principal (the original amount borrowed) and interest. An amortization schedule displays, in a table format, the amount of principal and interest included with each payment, along with the remaining loan balance.

Appraisal – The estimated value of a property based on a qualified appraiser's written analysis. Banks typically require appraisals before issuing loans to ensure the estimated value of the property adequately exceeds the amount borrowed.

Assessed value – The value of a property assigned by a governing authority to levy a tax or fee on the property owner.

Binder – *See sales contract.*

Buyer's agent – A real estate agent who represents the interests of homebuyers.

Closing costs – Incidental fees associated with completing real estate transactions, potentially including attorney's fees, credit report fees, document preparation fees, deed recording fees, appraisal fees, etc.

Contingencies – Particular conditions that must be met prior to closing a real estate transaction such as a home inspection (to ensure the home has no serious defects), a financing contingency (which releases a buyer from the sales contract if their loan falls through), or a contingency that a buyer must first sell their current home. In general, the fewer contingencies required of a seller, the stronger a buyer's negotiating position, in terms of getting the best price.

Earnest money – Also called a "good faith" deposit, these are funds held by a neutral party to demonstrate the buyer has serious interest in purchasing a property.

FHA loan – Loans insured by the Federal Housing Administration (FHA). With attractive financing rates and less stringent lending requirements than conventional mortgages, FHA loans are often appealing options for buyers with lower credit scores and/or smaller down payments. They do, however, require two types of mortgage insurance: an upfront premium and an annual premium, which is wrapped into monthly mortgage payments.

Fixed-rate mortgage – A conventional loan with a pre-determined (or "locked in") interest rate for the duration of the loan repayment period. They are traditionally 30 years in length but can be issued for 15 years, 10 years, or another duration.

Home inspection – A thorough professional examination (at the buyer's expense) that evaluates the structural and mechanical condition of a property (plumbing, foundation, roof, electrical, HVAC systems, etc.). This highly recommended step is a common contingency clause in real estate sales contracts. If the inspector identifies issues that may be expensive to remedy, these can be revisited with the seller before proceeding with the sale.

Listing – The printed (or digital) description of a property for sale. Listings may include details about the property, the home (number of bedrooms, baths, and featured rooms), other structures, the price, and photos.

Offer – A formal request to buy a home. *See sales contract.*

Points – Prepaid interest on a loan, equal to one percent of the loan amount. The advantage of paying points up front is that a lower interest rate can be secured for the lifetime of the loan. This may be a good deal if a buyer plans to stay in the home for many years (so the long-term interest savings outweigh the initial cost in points).

Pre-approval (loan) – A lender's written guarantee to grant a loan up to a specified amount (subject to receiving full documentation). Pre-approval for a loan can strengthen a buyer's negotiating position with a seller.

Pre-qualification – Less "official" than a mortgage pre-approval, banks offer (at no cost or obligation) pre-qualifications to estimate the amount a buyer may be able to borrow. It is often used early in a buyer's search to help determine a reasonable price range.

Private mortgage insurance (PMI) – A monthly insurance payment that may be required if a buyer's down payment is less than 20 percent of the home's purchase price. It protects lenders against loss if a borrower defaults.

Sales contract – A legal agreement between a buyer and seller to purchase real estate, for a specified price and terms, for a limited time period (also called a purchase agreement or a binder). When initially presented to a seller, this document is often called a purchase offer. Once the seller accepts (or the buyer accepts the seller's counter offer), it becomes a legally binding sales contract.

Seller's agent – The real estate agent who represents the seller of a piece of property. Their job is to act in the best interests of the seller, marketing their home to potential buyers and negotiating on the seller's behalf.

Title insurance – This type of insurance is acquired to protect against any unknown liens or debts that may be placed against the property. Before issuing title insurance, public records are searched to ensure that the current owner has legal rights to the title as well as the legal ability to sell the home and that no liens are held against the property. (*Google reference*)

WRITE THE DEFINITIONS OF THE FOLLOWING WORDS

- Advertising

  _____

  _____

- Cash Value

  _____

  _____

- Commission

  _____

  _____

- Copyright

  _____

  _____

- Disruption

  _____

  _____

- Distractions

  _____

  _____

- Employee

  _____

  _____

- Employer

  _____

  _____

- Entrepreneur

  _____

  _____

- Equity

  _____

  _____

- Expand

  _____

  _____

- Flipping

  _____

  _____

- Foreclosure

  _____

  _____

- For Sale

  _____

  _____

- For Sale By Owner

  _____

  _____

- Franchise

  _____

  _____

- Gentrification

  _____

  _____

- Idea

  _____

  _____

- Intellectual Property

  _____

  _____

- Investor

  _____

  _____

- Lease

  _____

  _____

- Manufacture

  _____

  _____

- Marketing

  _____

  _____

- Negotiate

  _____

  _____

- Owner Financing

  _____

  _____

- Ownership

  _____

  _____

- Partnership

  _____

  _____

- Patent

  _____

  _____

- Percentage

  _____

  _____

- Power Moves

  _____

  _____

- Redevelopment

  _____

  _____

- Rent

  _____

  _____

- Sit and Hold

  _____

  _____

- Small Business

  _____

  _____

- Trademark

  _____

  _____

REASONS TO PURCHASE REAL ESTATE

I had a chance to sit down with a lot of famous people in this industry to coach them and teach them a few things. But that's enough about me, let's talk about what is going on with the economy.

You have Donald Trump currently in office and a lot of people are talking about all the things he's doing but the reality is we need to be talking about the things that he's owned. What is his business? He is a construction magnate, a multi-billionaire. He has high rise buildings and golf courses all around the world with his name on it. So as you can see in the image to the left, he is digging with the gold shovels; digging in the dirt, breaking ground.

I had to go check things out for myself because I've been all around the country and noticed that Donald Trump's hotels are in most major states, and everywhere those hotels are, they are in very expensive areas.

QUESTIONS

- What was Trump's business before he became the president?

- Did you notice a lot of construction after he became president?

A lot of areas are being gentrified. So what's going on around the country? There is a construction boom that most people don't realize.

This is downtown everywhere, in every major city in America. There is a construction boom in all major big cities around America. What's going on? People who don't own real estate call it *gentrification*. People who do own real estate call it *redevelopment*, but if you don't know the game, what does it mean? It means you're going to get kicked out of the neighborhood; kicked out of your home, kicked out of the city; and you will eventually be kicked out of the country if you don't own a piece of land, a house, a commercial building, or some type of real estate that you can live in. So how did gentrification get started? Let's think about this. (This is New York City.)

QUESTIONS

- Can you identify new construction when you see it or do you know what to look for?

- Do you know the people that are building high rises in your city?

LET THE GAMES BEGIN

ɔ

.ie housing interest rate was 12.5% in Compton, California, where up. A small home was only worth $100,000 in 1985. By the time we co the year 2000 that house was worth $190,000. By the time we got to ∠006, the same little small house in Compton was worth $490,000.

Why was that? Because they got smart and decided to monetize the real estate. They decided to drop the interest rate from 12.5% down to 2% and by the time they did that, that real estate became really expensive. I hear a lot of people to this day say, "Oh, we better hurry up and buy a house because I heard the interest rates are going back up." You don't want to pray that they stay low, you want to pray they go up because that's what's going to make the houses more affordable where people can afford to live.

QUESTIONS

- What was the interest rate for California in 1985?

- What happened to housing prices when the interest rates went down?

Palos Verdes, California, rated the most beautiful city in America in 2016; a billion dollar area. A lot of millionaires live out there. A lot of people think that there is an invisible brick wall around it saying you can't afford it. People think houses there are always $2,000,000 and higher. But in 1985 a house that was 4,000 square feet looking at the ocean was only worth $340,000. But because they dropped the interest rate by the year 2000, that house was worth $890,000. By the year 2006 that house was worth over $2,000,000.

So what a lot of people have to realize is that if some people have paid their houses off at $340,000, just because the market went up and it says it's worth $2,000,000, you have to remember the house is already paid for. You may have qualified to buy a house at $2,000,000 or you have $2,000,000 that you can afford to buy a house with, that doesn't mean you should pay $2,000.000. You only pay the asking price in a seller's market. Then you may ask yourself, "What's next?" After you find out how much the house costs, the next question should be, "How long has it been on the market?" And the next question after that should be, "Well, if it's been on the market for over a year, how much do they owe on it? Oh, they only paid $340,000 and it's paid off or they paid $160,000." Well, if it has been on the market for over a year that means nobody really wants it. So why would you offer them $2,000,000? Now, it's time to negotiate.

"If it's paid off, how about I offer you $400,000 - $500,000 because nobody obviously wants it since it has been on the market for over a year or even 3 - 5 months." We are going to have to learn how to use that word called "negotiate." So, if you don't know how, call me and I'll show you how or I'll do it for you. This is one of the things that I can do as your coach.

QUESTIONS

- What cities are doing major construction?

- When mortgage rates go down, is that a good or bad thing?

POWER MOVES WITH WARREN BUFFET

There is a game to monetize real estate around the world, so let's talk about this. In 2003 Warren Buffet bought Clayton Homes for $1.7 billion dollars. In 2005 he invested in Forest River, which was America's best RV company. Clayton Homes was America's largest home builders and the biggest construction company. In 2008 Warren Buffett loaned Bank of America $5 billion dollars, which Bank of America had 30% of America's home mortgages on its books. And in 2008 Bank of America bought out Country Wide, which owned 60% of America's home mortgages. They paid $4.1 billion dollars. So now guess who owns 90% of America's home mortgages? Who owns millions of foreclosed homes? The one and only, Mr. Warren Buffet.

So all of a sudden in 2012, Mr. Warren Buffet bought out Prudential Real Estate office and changed the name to Berkshire Hathaway and they had over $150 billion dollars in houses that were on their books. He opened up one of the biggest Solar Farms in Lancaster, California, which for some weird reason Lancaster, California, and Mojave, California, became the Solar Farm capitals. There were solar panels everywhere driving the cost of land through the roof. So what are they doing around America? They are destroying home ownership for the lower classes by doing these moves they're making. They are removing home ownership and changing the whole paradigm to where people earning less than $150,000 are going to be renters.

QUESTIONS

- What billionaire started buying up everything that has to do with real estate?

- How much money did Warren Buffett loan Bank of America to buy Countrywide?

- What's the most expensive state to live in America?

So how were they going about it? They are building stadiums.

WHEN IS A GOOD TIME TO INVEST IN LAS VEGAS, NEVADA?

In Las Vegas, Nevada, there's a football stadium. It seems like every major city is getting a football stadium. So what is that doing? California's real estate is affecting America's real estate and a lot of people don't understand why. California's prices have become so high that it's pushing people out of California and they are moving all around the country. A lot of people are seeking more affordable ways and places to live, and even worse is a lot of people who graduate from college can't find a job in California or the states that they grew up, or growing up in. So they are being gentrified.

QUESTIONS

- How much did it cost to build the Raiders new stadium in Las Vegas?

- What happens to housing costs when people build new stadiums?

What is happening now is that you seem to have so many people moving to Georgia and now it's not affordable anymore. You used to be able to buy a nice beautiful home in Georgia for $5,000, $10,000, and $12,000, but now you have a lot of houses in Georgia going for $200,000 and still showing no sign of letting up. They are still going up. You have houses as much as $500,000; so it's pushing some people out of Georgia to Florida, South Carolina, North Carolina Virginia, etc.

Now the second wealthiest real estate place is New York and it's going up through the roof as well because you have a lot of people who have $1.2 million dollar apartments. If you don't own a house you don't have a backyard, so that's making people leave New York and they are moving to California because they have money as well. But they are also moving too. A lot of those people are moving to Georgia and now Georgia is the #1 destination in America for people who are being gentrified and that is why Georgia's prices are changing. Overall, the real estate prices in Georgia are going up and that is a great place to invest your money in right now.

QUESTIONS

- Why are people moving out of California?

- What's the #1 state African Americans are moving to in America?

WHEN IS IT A GOOD TIME TO INVEST IN LOS ANGELES?

How is California's real estate affecting the rest of the country and what does it have to do with you? For example, a house like this has 3 bedrooms, 2 bathrooms, and 1,570 square feet is going for $715,000. Keep in mind the people who purchased a house like this paid about $12,000 - $20,000 back in those days. Now it's worth $715,000. So what's going on? We have a lot of people around the country calling their family members saying, "Hey these people keep knocking on my door asking me if I want to sell my house?" So now you have family members saying, "Oh my Gosh they are going to offer you $715,000 for the house? Let me tell you what you can get for that kind of money." And we have a problem with that because people are misleading their family members because they don't know what to do with the money. So what are they doing?

WHEN IS IT A GOOD TIME TO INVEST IN GEORGIA?

They are telling them, "Hey come over here to Georgia," for example. "For $680,000 you can buy 6 bedrooms, 6 bathrooms, plus 7,500 square feet and then you can go buy a range rover and then go look for a job." I mean that's the kind of advice we give our family members with their $715,000 cash. "Oh, and by the way we can have Thanksgiving at your house; we can have Christmas parties; we can have everything at your house because you are going to have enough room, You are going to be the big dog. Everybody is going to love you. You can be the matriarch or patriarch because you will have the biggest house." Whomever has the biggest house wins, in the family world. That doesn't make any sense. We have to stop misleading our families to stop wasting money buying a big giant house when it's probably only 2 or 3 people living in the house in the first place. Four people seem to be the max sometimes for people who own homes and are moving out of state.

When educating our family and friends when moving, what are we telling people? "Oh, you can buy a big home for that money." "Oh, you can buy a house and pick your own colors, too." So those are the types of things that we tell our family members to get them to move to the state of Georgia and all these other places. What we should be telling them is, "You can build a nice house 3,000 square feet because that is double and triple the size of what you already have. Stop buying houses and start building houses, it's much more affordable. It's cheaper to build than it is to buy.

We can invest in some small businesses and some startups. Help them become successful by investing in so many businesses with that $715,000. Build some investment properties or commercial or residential houses. We have enough money to build our own housing tracks and sell them to other people because a lot of people are looking for houses right now. We can finance up and coming filmmakers and so many other businesses. We have a lot of businesses around America that are struggling. Then there's that one amazing store, and if they had $100,000 they could go open up about 2 or 3 more restaurants and you can partner with them. There are so many opportunities that we have, but we just need to know what to do with the money. That's where I come in helping out by teaching people what to do with the money and helping them to make their investments.

So when is it a great time to invest? When you have some money or you decide to sell your house. You can decide to pull some money out of your house. Where are the great areas and when is a great time to invest?

HISTORICAL REAL ESTATE INVESTMENTS

Let's talk about historical real estate investments. Purchase an apartment building or a house and rent it out. So what was the average rent going for back in the days - $500 a month? Over the years because of the high prices now, people are getting charged $1,500 a month for rent. That was a historical real estate investment after you purchased the home. Now, because of this major game changer in real estate called Airbnb, real estate has completely changed.

What's going on? Today's investment major game changer. Rent the individual rooms for $300 a week and that's $1,200 a month for 4 bedrooms; that's $4,800 a month. Keep in mind $300 a week is a low price. That's not including the bedrooms that have bathrooms with them because anything with a bathroom is going to get more money. Anything with a nice view is going to get more money. So now we are talking about $4,800 - $6,000 a month and coming from a little bitty house that you probably paid for $200,000. The real estate game has changed forever!

If you have ever played the game of Monopoly, you know that is is a board game that is designed for you to purchase property. I decided to use this analogy because I felt like this was a great example of teaching us great education back when we were young. We played this board game for hours. Most people do not realize that this game is the blueprint for real estate.

A lot of people participate through this format. What's going on here? It says, investors should start off by buying land and building houses, not buying one piece of land but buying multiple pieces of land. You can't buy one piece of land here and put a house on it. You have to buy all of the reds, all of the blues, greens, purples, yellows, oranges, whichever.

You can't buy three of the lots and then put one house. You have to build four houses on each lot. You have to go evenly with your four properties. So what does that imply? It's cheaper to build four houses than it is to build one house. So building houses is what we're supposed to be doing instead of buying houses, that's what Monopoly taught us as young adults.

What happens after we build those houses and they get old? Once you get all four houses on each lot, now you have twelve houses on all four; all three of your lots. Do you sell them because the real estate agents are knocking on your door saying, "I know you have twelve houses. We're willing to offer you millions of dollars for them." You say "no" because you can tear them down and build apartment buildings or hotels like Monopoly taught us to do.

QUESTIONS

- How many square feet was the mansion in McDonough, Georgia in this book?

- What business created a major disruption in real estate?

- What board game teaches children how to become investors?

HOME-BASED BUSINESSES ARE ON THE RISE

A lot of high rise buildings are going up all around America. Why is that? Because America has realized that through the internet, we can have a home-based business and set up at home and make a lot of money. People are making millions of dollars all around the world and the average everyday people are sitting at home coming up with online businesses becoming six-and-seven-figure-year earners.

So what did America realize? We want to get these guys out of the houses and give them more space. They need to come out of the house and a have business in their office. America is building office spaces all around the world, building high rise buildings and then renting out office space because there is a huge demand for it from these small businesses.

For example, now you have Regus, Roam, Wework, and so many other places who are opening up all around America and saying, "Hey, new business owner, you can open up your business here. We have office space for you. We'll rent you some office space. You'll have a view; you'll have all the latest technology to use, if you need it."

What is that called? I call that the new face of business; but guess what, when you get older and you decide to retire, you can't leave that building to your children. You can't leave that office space that you were renting at Regus to your children and you definitely can't protect your ideas sitting in this office space where they have it camera'd up and mic'd up recording all of your intellectual properties and waiting on you to come up with these new ideas.

QUESTION

- What is Wework, Roam, and Regus?

SECURING SMALL BUSINESS OWNERSHIP

This building right here is very important that we buy these right now while we can afford these because what people realize is that you can run a business downstairs and you can live in the apartment upstairs. You figure if you pay $150,000 - $200,000 for that building, your mortgage will only be $1,200 - $1,500 a month. For example, if you made a barber shop out of it, you can afford to charge ten dollars for a haircut because you live upstairs; you're saving yourself thousands of dollars a month and it'll reflect your prices.

Now they're trying to make all of America look like New York. They want to own the entire building, so now you can't buy the buildings at the bottom, you can only rent them. If you live above the buildings just like in New York at $600,000 for a house or apartment above, you pay $600,000 to live above these commercial buildings that will be $3,200 a month. Then you decide to rent one of these lower units at a minimum. It may cost $6,000 - $8,000 a month in California at least.

So what's it going to cost you for a haircut now if you're paying $3,200 for your mortgage and $6,000 - $8,000 to rent a business? You're going to be paying $50 - $60 for a haircut. This is why we've got to start buying these buildings because of ownership. If you don't own a business you're going to be paying higher prices because if you're down here at $6,000 - $8,000 a month today, by the time tomorrow gets here, you could be paying $10,000 -$12,000 a month for rent and those haircuts are going to keep going up.

I tell people to stay away from this, but then at the same time I tell people to invest and start buying buildings. Start renting commercial real estate. You rent the entire floor yourself for around $30,000 a month and then you rent out each unit and possibly make $200,000 - $300,000 a month all for your real estate by opening up your own Roam, Regus, or Wework. I see a lot of these places popping up all around the country. In fact, I visited a black owned business in Atlanta, Georgia. In Atlanta there is one business called "The Office Spot." But what's going on is that they're renting their office space making a killing charging people monthly fees and that's a great business. I love it. I love the model.

So what's going on in L.A. and around the country? People are buying the houses, the smaller houses, and tearing them down and building up bigger houses. So I tell people don't take the $600,000 or $700,000 for your house. Tear it down and build a bigger one that's worth $3,000,000 and $4,000,000. If your house is paid off you can use that same money that's in your house to tear it down and pay to buy some blueprints and build a bigger house worth more money. Then you have the option to sell it or you can keep it.

QUESTIONS

- Why are investors buying small houses and buildings to tear them down?

- Why are places like Wework and Regus making a lot of money?

LOST LEGACIES

Let's talk about people who get caught up in the conspiracy of real estate and the conspiracy that we create for our own selves. We have a lot of people who went to visit "grandma's" house. Grandma lived in Alabama, Louisiana, Mississippi, all of these little small states around the country where grandma owned these houses and had a little lake in the backyard or a big lake, but all you remember is that you got bit by a snake or you didn't like the big bugs that were flying around grandma's property or you had to run from an alligator when you were a kid. So you don't want anything to do with grandma and grandpa's house.

Now, grandma and grandpa passed away. They decided to leave you the house with the will and now you are thinking that you've got your college degree, you've moved on and got your new car, then moved on to LA or New York now. You don't want anything to do with the house in Georgia, Alabama, or wherever grandma and grandpa lived. Now you don't want that land, that 15 acres that grandma and grandpa left you. So you decide to sell it. This is a big problem in our community. People are selling grandma's and grandpa's house on that land.

So what do they do? They get on the phone, they Googled real estate and call up an investor and the investor's thinking, "You want to sell your grandparents' house because you live in New York or you live in LA or you're a city slicker. You don't want to go back home because it's too slow out there, because they have alligators? Oh really. Okay, well what do you want for it? Oh, you don't know what it's worth? You're not going to fly out here? You're not going to send "TJ" out there to investigate it? You're a city slicker, you don't have time to come out here and see it. You hadn't been here since you were 8 years old?"

What does the guy say, "Oh, since you're not going to come out, I'll tell you what, I'll see if I can help you out. You know you were right, it was snakes and all of that. Maybe I can offer you five grand or something like that." You're thinking like wow I don't have to pay the taxes anymore; I'm coming up. I don't have to send "TJ" out there to see if any new corporations moved in.

Next thing you know, you take the five grand. What does this guy do? He's excited; he got him another sucker now. He's checking out his 15 acres sitting on the lake. He calls his people and asks, "Hey what can we do with that 15 acres?" And he says you know that Google, and Amazon, and Mercedes Benz just moved in. They need some custom homes in the area, so that they can put their corporate execs in. They call their investor friends and make plans.

The next thing you know they put corporate housing up, mega-mansions to house corporate executives of the big corporations that recently just moved into the more affordable areas like your grandma's area because they realized it has become so expensive to live in the big cities. Now, they're moving out further bringing the jobs that's bringing all the employees out. They're not trying to commute, so they decide that they're going to move to the local areas. They take grandma's 15 acres, subdivide everything into half acres, and build some 10,000 square feet mansions on each lot and sell them for millions of dollars each. You stop by and ask, "What happened to all the snakes and alligators and all the big bugs?" You find out that they've been gentrified, too.

These guys dug the bottom of that little small lake out that you could barely fit your boat on and made it to where you can bring a mega yacht through there. So now this area is one of the most lucrative areas in that area and you can't even afford to go through there anymore because if you do, the police might stop you.

QUESTIONS

- Why is it important not to sell your parents or grandparents' home when they pass away and leave it to the family?

- What is subdivide?

THE HIGH COST OF LIVING IN AMERICA

Live in Los Angeles and you have to make over $115,000 just to live in L.A. To live in New York, you have to make over $99,000, but most college students don't even make $80,000. Therefore, college is not always the best solution. So let's think about it. Why is college not the only solution? Baby boomers went to college for free or up to $50,000 to go to college. Homes cost the baby boomers $12,000 - $20,000 back in those days. Now those same homes are worth $400,000 - $800,000, and most people don't realize what happens when you buy a home and all of a sudden you have equity coming in.

Well, you start getting credit cards worth a lot of money. So what happens when you find a job? They get hired from jobs paying from $40,000 to $80,000. Keep in mind that a lot of people didn't have to go college back in the day. They just walked into General Motors and got hired

because General Motors needed help and they trained on the spot.
"Most people are credit and equity rich but cash poor."

Let's talk about generation Z, post-millennials born in the mid 90's and the 2000's. College is $160,000 - $300,000 for those young people and the homes are $400,000 - $800,000, which keep in mind, you have to make over $100,000 to afford some of these homes. So what's going on? Credit cards are maxed out and they're paying for books, etc., trying to get home to mom and pop because they're moving further out now.

Jobs are paying $40,000 - $80,000 annually. So what's going on with that? We've been bamboozled. Not "we" but most people are being bamboozled. Back in the baby-boomer days mom and dad made $40,000 - $80,000. Now in 2018, the young people are making $40,000 - $60,000. So anyone making over $60,000 is exceptional because that makes up a smaller percentage. Anyone making over $80,000 is a definitely above an exceptional person in the working world, financially speaking. So what's going on in the working world? Ninety-five percent of our college students are unemployed or underemployed.

So what's happening? We are raising the first generation of non-home owners, and this is dangerous. Because if you don't own a home that you live in then you are in a situation to be gentrified and pushed out. This is so important to start buying houses right now while we can afford to buy them. So you may not like the neighborhood that you're thinking about moving to. But if you can afford it, you better buy it now. If you think you're going to wait until next year so you can get your tax return, or you can have some things knocked off your credit by then, or you can make some more money by then, you might not be able to afford to buy the house that you didn't want to live in, if you keep waiting.

QUESTION

- Why is it important to know how much will it cost to live before you go to school and take your major?

COLLEGE ALONE IS NOT ENOUGH

A simple change is for college students to primarily take classes that they can come home and start businesses within their communities. That way they can create jobs and hire local residents. By starting businesses, they can purchase the commercial real estate in the areas that they grew up in.

Being on a college campus and surrounding yourself with a lot of smart people, this can open up an amazing opportunity within the business world by having different relationships. Every business needs an attorney, an accountant, someone in marketing, a good political friend, a celebrity to endorse the business, and a scientist to create the new products. There are endless opportunities that can come from college with just a little coaching from me and a few tweaks; then we can turn this entire system around. Higher education means the ability to create jobs within the community.

Gentrification is created due to people who don't come back to their communities and create jobs within their communities. So what happens? If you're not coming home from college and creating jobs in the community you grew up in, you're going to open the door for someone else to come in. That's why you go into a black community and you have large amounts of other people owning all of the businesses.

They are coming into these communities buying land and gentrifying them. So, now African Americans aren't able to afford to purchase the house that they once grew up in.

This building on Howard University's campus shows a sign of gentrification. With all of the homes being sold around the university, it's only a matter of time before other people buy up all of the homes and decide that they no longer have a need for African American universities around the country.

QUESTIONS

- What types of classes should be taught that will enable students to find jobs or start businesses that will allow them to generate $150,000 and higher?

- If non-African Americans enrol in large numbers to HBCUs, is this opening the door to gentrification in African American communities?

- If yes, please explain:

ADDIDTIONAL QUESTIONS

- Do you plan to attend an HBCU or any other educational institution?

- What field of study will you major in?

- Have you researched that field of study to ensure its longevity beyond a few years?

- Have you prepared yourself financially to pay for your education via scholarships, grants, loans, etc.?

Student Loan vs. Jail Time

The government has recently attached the marshal's office to the student loan department. They are arresting people who received student loans over 20 years ago and are working their way up to the current debtors that haven't paid their loans off. Unfortunately, this is not common knowledge of what's going on in the educational system. People don't realize that they're signing up to one day possibly go to prison due to non-payment or lack of being able to pay.

So let's look at good jobs. We all tell our children get good jobs. So you ask yourself, do you want a good job? A good job is basically something that pays $60,000 - $80,000; anything over $100,000 is a great job. Now, do you want your children to have jobs or do you want them to be wealthy? Did you know that the majority of business owners are ex-felons? Less than 2% of them are college students. So what's going on, what happened to all of these college students? Where are they working now? They're all driving Uber, Lift, etc. But now we have electric cars coming out which are going to kill all of the Uber and Lyft drivers. We have electric diesels now. As of February 2018, the new law passed in certain states for driverless vehicles and we don't need anyone to be behind the wheel anymore.

ELECTRIC JOB KILLERS

Elon Musk makes Tesla. Tesla trucks will drive themselves. You can literally go to your app, tell your car to meet you outside of the bar or meet you outside of the club and pick you up at 2:00, and in 30 minutes your car will start itself up and head to pick you up. Seven million drivers as of January 2018 are driving for Uber. Lyft has over 700,000 as of August, 2017. There are 8,700,000 truck drivers all around America. All of these people are going to be out of work very soon when that electric car law passes all over the country.

QUESTIONS

- Why don't some people like electric cars?

- What industries will lose jobs once electric cars are fully legal?

DISTRACTIONS

We have to watch out for the distractions. What kind of distractions are there? We have cryptocurrencies and we have stocks. A lot of people think that cryptocurrency is the answer but whenever you have a good situation in America coming up and people are making money in real estate, you're going to have a distraction out there. Cryptocurrencies are a distraction and the stock market is a distraction. People in the stock market are taking your money while you invest in them. They're taking your money and buying up your houses and gentrifying your neighborhoods. So I tell people to start your own business and invest in them.

Gold is a distraction, MLM's are distractions, and marijuana laws are a distraction. So we have to remember: **"The powers that be will give up a million-dollar game to keep you distracted from the multi trillion-dollar game,"** that's my original quote. Why is that so? I never said that cryptocurrencies, stocks, gold, and marijuana will not make you any money, but these are pennies on the table compared to land development. Every neighborhood in every major city is going through mass gentrification and mass construction. So if you are a land developer, you are en route to becoming a billionaire.

During the housing crash, back in the days, some people who owned homes filed lawsuits to save their houses from the bank that was trying to take them because they signed documents getting bad loans. So what happened? Some people decided to fight to save their houses, hiring attorneys for $15,000, $30,000 and then 10 years later you ask them, "Did you win your house back from the bank?" You say, "No, I had to hire another attorney to sue the last attorney that I paid to sue the bank."

Years later black people are just now getting modifications or a lot of them have lost their homes even though they paid $15,000, $30,000, $50,000 fighting to save their homes. What did other people do? Some people decided to walk away from their homes and purchase themselves RVs and moved into their RVs on RV parks and saved money.

They took the hit on their credit and let the creditors take the houses and messed up their credit. But guess what? Ten (10) years later, 7 years later, the foreclosure fell off of their credit, they filed bankruptcy, their credit is back flawless and perfect and the world is a wonderful place for them. Now they're back in the same but better neighborhoods buying bigger and better houses for cheaper prices, putting more money down because they've saved money over the years.

QUESTIONS

- Why is knowing the salary you need to buy a home important?

- What is the stock market?

- What is the stock market buying up all around America?

- What is a distraction?

WHEN IS A GOOD TIME TO INVEST IN
OXON HILL/NATIONAL HARBOR, MARYLAND?

I visited Oxon Hill, MD, and National Harbor, MD. However, PG County (Prince George), MD, is actually the third wealthiest county in America for African Americans, but guess what? PG County, MD, is being gentrified. They built a brand new MGM Hotel sitting right here.

This particular area is very expensive. This small black community of people is being gentrified like crazy! Prices are going up so fast due to all of the major corporations moving in; such as, the casinos and the big businesses. This is definitely a place to invest in.

Households like these high-end townhomes are being built in an African American neighborhood that were once owned at a price of $150,000 for a home near the water. Now, those older homes are being torn down and these 4-story townhomes are being built up all over the place at a price of $750,000 and up.

Keep in mind you can never really own a townhouse, condo, or a home in a neighborhood that's charging people association fees or anything other than your mortgage payment and insurance. You can't have a garden on your property. Also, when purchasing a 3+ story home, always remember older people can't climb stairs safely.

QUESTIONS

- Do you see townhomes being built or are they already in your area?

- What else can't you do with a townhome?

Whenever you see a strip mall like Tanger Outlet pop up, that location is an up-and-coming area and that's a great time to invest. We should be buying the real estate up around all of these places. If you find out there is a Tanger Outlet popping up or if you find out there is a casino popping up, that's a great area to invest in. So when is a great time to invest?

QUESTIONS

1. What kind of business would you put in or around an outlet where thousands of people shop monthly?

2. How much do you think it would cost you to build an outlet in your area?

MONETIZING GENTRIFICATION IN WASHINGTON, D.C.

I visited Washington, DC, and California, it's the same thing over and over; black neighborhoods are being gentrified. You have the hood where the black folks live and then you walk across the street and you notice that they have brand new high rise buildings going up all over the city.

All over the community you literally can walk two blocks over from the hood where people are getting shot at and go over to the Burberry, Gucci, or Louis Vuitton buildings and buy some high-end products from the brand new strip malls and buildings that they built in the hood. Never before have I seen anything like this in the hood. This shows me what's going to happen in Watts, California, and the rest of America.

A great way to monetize gentrification is to pay close attention to what's going on within the Black communities by attending community meetings. You will find out who is doing the gentrifying and which corporations are moving into your communities. By knowing who's coming, it opens the door for investments. It opens the door for very lucrative investments.

QUESTIONS

- What's a sign of gentrification?

- Why do you think they hold these meetings when most people are at work?

- Do you know how valuable it is knowing which major corporations are buying up your neighborhood?

- What could you do with this information?

MONETIZING GENTRIFICATION IN GEORGIA
AUTOMOTIVE POWER MOVES

Well, this is where Mercedes Benz decided they were going to open their Corporate Headquarters in the middle of nowhere inside of Georgia. So this lets me know that at that particular moment was a great time to invest and buy land from the real estate developers from across the street, around the corner, or anywhere in the area. Think about it, all of those corporate executives and all of those employees are going to need houses to live in or they are going to want to buy some coffee.

There are a lot of business opportunities, a lot of opportunities to make money and investments. So, when Mercedes Benz moves in, you are supposed to stake the neighborhood out and figure out what's missing. People are going to want to catch the bus to work; they are going to look at getting off the bus or getting off the subway or whatever it is and they are going to be looking for some coffee and snacks. They are going to be looking to plug in their laptop after work; they are going to want to go to a club, etc. There are so many opportunities to build businesses around the corner across the street from the Mercedes Benz Corporate offices and so many other corporate offices like that.

QUESTION

- What state did Mercedes Benz move its corporate headquarters to in 2018?

MAJOR CORPORATIONS RELOCATING THEIR CORPORATE HEADQUARTERS ALL OVER THE COUNTRY OVER THE PAST DECADE

Major corporations are moving their corporate headquarters all over the country creating large amounts of wealth for themselves and for everyday people who have an eye of an investor. So Mercedes-Benz opened its headquarters in 2018 and purchased 900 acres of land. And built a brand new corporate headquarters as well as built a 33,000 sq. ft. shopping center with a large number of homes to accommodate its employees as well as others who like to live in the area. Corporations have gotten smarter over the decades. They realize that it's more affordable and very lucrative to move to a smaller city.

The strategy for corporations has been to move to small towns and buy affordable properties to build their corporate offices and facilities, etc., once their prices increase due to the small businesses moving near to accommodate the major corporations' employees. This pushes the cost of living in the area. For example, most of the time the city will give them the land for free and when they sell the properties and they're ready to move, the properties will be worth hundreds of millions if not billions of dollars.

So when they move to the new areas now getting more property to build, this time corporations are buying buildings, shopping malls, and hundreds, if not thousands of homes in the area and licensing out the land to other businesses like gas stations, etc., that want to be in the area. To buy on the outskirts of the land that's owned by the corporations can be very lucrative or to buy early in the first phase of the housing track that's being built by the corporation seeing that after a few years after everything has been completed, mass amounts of wealth can be earned. For example if purchasing a house in the early stages for $350,000 two years later after the plan has been completed, the house can be easily worth $1.2 Million and up. Also a person renting one of the store fronts in the brand new shopping center can create large amounts of wealth in the process by selling something as small as smoothies and salads to sandwiches.

One small catch is the corporations have gotten smarter and are charging monthly association fees for everything within reach. So the homes will have to pay $200 per month and higher for any properties you purchase from the corporations.

To sum it up, one of the best ways to monetize gentrification is to pay attention when hearing that cities are going to donate land to build facilities in a certain city or state and buying some land around that area or starting a business anywhere near that area is a huge way to monetize gentrification.

Mercedes-Benz also came in and built this $1.2 Billion dollar stadium in Atlanta, Georgia, creating amazing investment opportunities around the stadium. When stadiums go up, a parking lot, sandwich shops, smoothie shops, restaurants, homes rented out to Airbnbs, anything pertaining to sports is going to make large amounts of money being around the stadium. So this is definitely a win-win situation for investors looking for a great investment.

QUESTIONS

- How much money do you think Mercedes Benz made selling their old corporate headquarters?

- How much money do you think Mercedes Benz paid for the land they built their new office on?

When Tyler Perry's Studios opened up in the middle of the hood, in East Point, Georgia, that became another great time to invest. Again, buying up all of the real estate. You don't have to buy it all up. If you can buy a house, a regular house, just buy it and sit on it and because it's near Tyler Perry's Studios, you are going to make money from that. So, of course, I had to go out there and check that out for myself and I saw a lot of very affordable neighborhoods that are in the hood. Where are we now? Atlanta, Georgia, has become in 2018 the biggest film state in America, in the world. They beat out Hollywood, CA. The first time ever in history. Georgia is definitely the place to invest!

QUESTION

- Why do you think the housing prices are going up near the new studio?

I was in Lithonia, Georgia, and I ran across a closed down Taco Bell that was only on sale for $250,000 for the entire building, parking lot and everything. I thought "What a great opportunity for a business owner to buy this and put their business in." We have to start buying buildings to run our businesses. So in California, you would never find a building this cheap because the prices are so expensive. For this much land, this building in California will be worth about $2 million dollars, so because of all the people from California moving to Atlanta, Lithonia, etc., this building one day will be worth $2 million dollars or more. So why not invest in that and put your business in it?

So while you're selling food, vegan food or whatever it is you make, you would be making money on your business selling food and then also you would be making money on your real estate because you own the building.

These old buildings sat vacant and for sale for years and are now being restored at an affordable cost and sold or available for rent for top dollar. Commercial office space on the bottom and residential apartments on the top. A huge way to create wealth is by living in one of your residential apartments upstairs and running your business downstairs.

With all the new updates available, restoring these block buildings have become a lucrative industry. Buy low and sell un-affordably high. With small investments for upgrades, it can equal massive profits in the end. After watching these two building being restored from eyesores to most sought after, I have a total new perception for what you can do to an old building. It's very important that we buy these buildings while they're still affordable. They play a pivotal role in African American business ownership.

Atlanta, Georgia

I always look for houses with dumpsters in the front yard. It lets me know that investors are at work. Dumpsters are the start of homes being renovated and potential wealth being created.

A house like this will discourage you but can be proven to be a very affordable restoration job if priced right. I will find a seasoned investor before purchasing a property like this because some of the houses like this one may be too far gone and should be torn down and rebuilt.

If I see brand new houses in neighborhoods where there are trash bins, it's a guarantee that that community is being gentrified when old houses are being torn down and new houses are being built.

Building basic simple track houses can create large amounts of wealth and plenty of jobs during a seller's market when existing inventories are low. It's always good to buy in the first phase because by the time they get to the fifth or sixth phase, you could have possibly amassed a small fortune in equity.

QUESTIONS

- Why does building houses create large amounts of wealth?

- Why do you think building houses creates so many jobs?

- What type of trades do you think people should have to build houses?

Because of gentrification around the country, people are flocking to affordable cities buying and renting affordable housing. Historically, people don't inquire about the cost of purchasing a commercial building, but it is very affordable seeing that this 5-plex shopping center in Georgia was on the market at only $150,000. This is a great investment. It allows a small business owner to run their business basically free while the other four tenants pay the mortgage and allow the owner to put a hefty profit in their pocket. Seeing that the average rent is at an all-time high where you can get from $800 - $1,200 a month each and even allow a family member or friend the opportunity to start an up and coming business and create possible partnerships, this is a win-win opportunity.

QUESTIONS

- What type of businesses do you normally see in strip malls?

- What type of business would you open up in a strip mall?

- Why do think owning a strip mall is a good investment?

WALL STREET: DESTROYERS OF GENERATIONAL WEALTH

Wall Street has decided with Warren Buffett and Berkshire Hathaway that they're going to start buying all the homes around the country and rent them out. Want to know why? Because of companies like Airbnb see that the cost of living is going up and the cost of real estate is going up. For those people who are thinking "Oh, the market is about crash I'll wait, I'll wait, I'll wait." I can coach you and teach you the game and you'll see that this shows you that real estate isn't going anywhere now. It's not going to crash anytime soon because so many people have changed the game. They are removing home ownership and creating wealth by renting out all the properties. So let's discuss that. I decided to go deeper and check things out for myself.

So I took a trip out to New York myself to check things out. I decided to go to Wall Street where the players play. This is me on Wall Street checking out where all these major corporations around America are buying up all of the real estate in America and renting it out, putting it on Airbnb, for example, because they want to reap the benefits of the high cost of living in America.

QUESTIONS

- Why do you think Wall Street decided to start buying up all of the real estate around the country?

- Do you know what was the first stock publicly traded on Wall Street?

MONETIZING GENTRIFICATION IN NEW YORK

So, here I'm in New York and a lot of people see these high rise buildings and say, "Oh they are too expensive." That building costs $6,300,000. A lot of people see stuff like that and don't even think about it because they know that they can't afford it. That's an invisible brick wall around that building that says you can't afford it. So most people don't know what that building costs, but I invest in real estate and I deal with stuff like this all of the time.

That building is 46,745 square feet; so what I look at is that you may not be able to buy that building by yourself but as a community, you definitely can. Can you imagine having 20 people going in to buy this house, this building? That building has to be about 20 stories tall probably. So what happens then? Everybody pulls their money together and buys this building and all the people who invested in the building can move in the building and rent the other half of the building out and make millions of dollars off of this building. It's like teamwork makes the dream work. That's how simple it is. So don't ever let big prices scare you. If we can come together and build a megachurch, we can definitely come together and buy a high rise building and everybody can get their own place and reap the benefits from it.

QUESTIONS

- Why do people buy commercial buildings?

- What's the most important investment you can ever make?

- Why is it important to have your own business?

- Why is it not a good idea to rent houses or commercial real estate anymore?

- Why is buying homes and commercial real estate creating wealth for everyday people?

So this is me sitting in what used to be the World Trade Center where so many people passed away. What's the relevance of this place? This was a horrific place where a lot of people lost their lives and where planes crashed into the buildings. It has become a huge tourist destination for people around the world. A lot of people want to come see what happened.

So what did they do? They got smart and decided to make it a shopping mall because you have a lot of people that come here to buy. They want to buy trinkets to take back home. This has to be some of the most expensive real estate or expensive area to shop because it's a huge demand. So many people want to go see the World Trade Center where planes flew into and knocked the buildings down.

They were doing some construction work in front of the famous Stock Exchange building. Some people decided they're going to buy up all the real estate around America and we are going to rent it out. We're going to remove home ownership.

Buying these 3-story houses in New York tearing them down and building them larger can prove to be very lucrative especially renting them out for generational wealth.

QUESTIONS

- Why do you think New York wants investors to build larger buildings?

- What is generational wealth?

MONETIZING GENTRIFICATION IN PENNSYLVANIA

This is me in Philadelphia, Pennsylvania, and someone decided that they were not going to sell out to gentrification. They were not going to leave; they were not going to take the money. So this house was probably worth $150,000; three-story house, 3 bedrooms, 4 bedrooms, whatever the case may be. The owner decided not to sell the property. They offered $500,000, then they offered $700,000, but the owner kept turning them down.

So they eventually decided to build around it. Everybody talked bad about this owner. They laughed and thought the owner was stupid. But sure enough after the owner kept that house and did not sell out, they just started building around it.

They changed the zoning making it to where buildings could go up taller. So as you can see 1, 2, 3, 4, 5 stories where it was barely 3 stories tall.

Now you can go up taller than 5 stories. This house was probably worth about what they started off offering between $300,000 - $400,000 to get the owner out of there. Now this place is worth about $2 million dollars easily. He can tear this place down himself and build his own building up just like the other ones here and do very well with it. That is a great investment. I'm actually proud of this owner. All the owner has to do is come in and tear that building down and put up a brand new building where the owner can go up taller now just like they did. Put up a 5-story building and probably get like 30 units inside of there and get top dollar for them. The options are endless.

QUESTIONS

- Why do you think investors are tearing down small houses and building larger apartment buildings, townhouses, and condos?

- Why do you think people sell their houses early and lose so much money instead of waiting and creating generational wealth?

- Why are people starting to buy land?

Here in Philadelphia, Pennsylvania, developers are purchasing entire blocks of houses and building apartment houses, condos, and new homes. Creating hundreds of jobs for each project; such as plumbers, electricians, dry wallers, scalfolders, painters, masons, etc. None of which require a college degree but all earn an easy six figures a year during the housing boom with minor training.

In this picture, developers have developed half of the block, which means the other half of the block didn't want to sell (homeowners or apartment building owners). This is a great investment. I would always inquire to these owners if they were interested in selling due to the previous development increasing the equity in their properties. If you can convince them to sell affordably, you can tear down and build the rest of the block brand new making more money than the first half of the block with all of its construction since they've already increased the property value on the block. This opens the door for you to charge more money per house.

QUESTIONS

- Why do you think it's very important to keep in contact with older people who own homes in highly gentrified areas?

- Why do you think it's important to buy an older home surrounded by new homes?

- Why do you think other races of people move into predominately African American communities and are willing to pay more money?

Missed opportunities for wealth such as broken down brick walls rigged up to work is a huge example of showing the effects of trades not being in schools. Not being able to repair a simple brick wall, not knowing how to fix a broken window, loose roof shingles or tiles, falling down trees, overgrown or dead grass from a broken water sprinkler, a stopped-up toilet and sink, raised or ripped vinyl flooring, rotted out rear decks or broken front steps and many more household issues are all examples of the trades being removed from the schools. This plays a major part during gentrification seeing that too many household repairs drastically decrease home values. Each of these repairs can be completed for less than $100 if you have the simple trades that should be taught in every high school.

QUESTION

- Why do you think learning a trade like building brick walls can create generational wealth?

Young individuals are making money painting murals on the side of buildings all around the country. With even more buildings going up the door is open for more wealth to be created painting murals. One simply needs to get in contact with the owner and ask would they like a custom mural drawn on the side of the building to keep the graffiti down.

Philadelphia, Pennsylvania

Philadelphia, Pennsylvania

QUESTIONS

- Why do you think cities pay large amounts of money to paint murals on walls of buildings?

- Why do you think it's important to learn how to draw while you're young?

MONETIZING GENTRIFICATION IN CALIFORNIA

LEAVING MONEY ON THE TABLE!

Let's talk about people selling their houses too fast. This is Venice Beach, California.

So what's happening here is that the people who owned this house sold out too early. They probably sold out at $500,000 - $600,000 because they only paid $12,000 - $25,000 for their house. So, now as you can see it has 2 bedrooms, 2 bathrooms, 1,056 square feet and it's worth $2 million dollars "as is" right now. Think about if you sold this house for $500,000 and laughed all the way to the bank thinking you did very well. But then you realized two (2) years later you drove by and that same house that has never been painted, never even changed the water hose, didn't do anything, yet is worth $2 million dollars and the whole neighborhood looks completely different.

QUESTIONS

- Have you noticed any houses around your neighborhood that need to be refurbished?

- If you could purchase that building now, what would you do with it?

- Have you ever considered researching the property in your community to see what the status is in general?

BEWARE OF SCAMMERS OUT TO TAKE YOUR MONEY!

All across the country there are wealth conferences, property flipping conferences, real estate expos, etcetera, designed to take people's newly acquired wealth in their homes called equity. Make sure you research their business entirely before you invest your money and time. Don't become their next target!

QUESTION

- Why do you think African Americans don't show up in large numbers to learn how to become investors and buy real estate?

An unfinished house is always a top choice when wanting to complete and rent out for personal use. Houses like this one can cost pennies on the dollar because developers dropped the ball and lost the homes back to the banks. That makes them very appealing.

California City, California

QUESTIONS

- Why do you think some real estate agents don't show African Americans houses like this?

- Why do you think these unfinished houses are so affordable?

The BEDFORD GROUP is a Black Owned Land Development company based out of Los Angeles, CA who's been building large scale luxury condos across the country for decades. They've always recognized the value in our communities and have been developing them for years. This is a large scale project about to be built overlooking Los Angeles which would be able to see the Pacific Ocean.

The Bedford Group land overlooking the Ocean and the city of Los Angeles, California.

The home of the new Los Angeles Rams and Clippers stadium is driving the real estate prices crazy in Inglewood, California. With the Olympics coming, Super Bowls, and so many other major events, Inglewood is going to become one of the wealthiest cities in America and I'd be very happy just to own a hotdog stand or anything across the street. With this large construction site, building hundreds of apartments renting only (nothing for sale) there are enough construction jobs for Inglewood, California, for years to come. It's predicted that by 2050 Los Angeles County will have 10 million new residents. So Los Angeles is under heavy construction to accommodate people. And also you will see a lot of new construction projects being built non-car friendly, meaning that less parking and less parking lots and plenty of Uber/Lyft pickup spots. Also, they're building for electric cars' charging stations to accommodate the rapid growth of electric cars.

INVESTMENTS AROUND CALIFORNIA

So, what's going on and why is real estate going up so expensive in California? For one, Donald Trump built one of the most expensive golf courses in Palos Verdes, California. Terrenea is one of the wealthiest resorts in Palos Verdes, California, to ever come to America.

Ferrari and Porsche opened up in Torrance, California, which was never before done. Ferrari was never there before. Porsche built a brand new facility. They also built a race track in Carson, California. Before you buy a brand new Porsche, you can test drive it first around the race track. *Sit and Sleep* opened up a huge building in Compton. UPS and Best Buy opened up huge warehouses in Compton making Compton some of their biggest distributors.

All of these trillionaires and billionaires are moving into the same area and they're driving real estate crazy. That's opening the door for people to create wealth by starting businesses and owning real estate.

QUESTIONS

- What happens to real estate prices when big corporations all move to the same area?

- Why do sport car companies open new stores around major corporations?

I'm in Rancho Palos Verdes Estates, California, deemed the most beautiful city in America. This is a 6,500 sq. ft. mansion overlooking Los Angeles county and the ocean. Someone purchased an existing smaller home on this property to tear it down to build a bigger house instead. This is a great investment seeing that the land that this lot sits on has an amazing view. People with large amounts of money love to have large homes with at least a 180 degree view and up on the oceans. Years ago, developers built small homes on view lots. Now all of the good lots seem to be taken; so it's a good strategy to buy up the small homes on the view lots and build larger homes. This is not just limited to California, but it's happening all over the world.

Even tourist destinations like Catalina Island are experiencing major construction to bring in more money for all of the new tourists. Again, you see a small house that was torn down and a larger house is being built overlooking the ocean and the state of California.

This small house has a 180 degree view overlooking Catalina Island, the ocean, and California's coastline. This investor is completely remodeling the home and not adding any more square footage to make it larger. You can also see where they cut into the brick wall and added 2 two-car garages. Personally, I've been monitoring these houses because I'm sure this investor is leaving money on the table by not tearing this house down and building a larger one. So, I've been watching houses like this as a possible future investment.

Restaurants in Catalina Island and all over the country are pulling down large profits from tourists simply selling bottled water and basic foods. Opening any type of restaurant around a hotel will bring in large profits.

This is a very small house on Catalina Island, California, being used as a rental property. The majority of the houses are very old on the island and can stand to be torn down and be built up to date. Tearing down the older houses and building brand new updated houses and renting them out is a great way to create large amounts of wealth. When people are traveling on vacation and renting the house, they almost always prefer a newer home. Those are the ones that are highly sought after. On Catalina Island a lot of the homes are outdated, which is an investor's heaven due to the large amount of rent being paid.

This is a fast-food restaurant popping up in Lancaster, California. Because of the high cost of hotel food, having a restaurant within walking distance of a hotel is a very lucrative business opportunity. A business can easily make seven figures a year just by selling bottled water and snacks.

Here in Newport Beach, California, buying the smallest house with a large lot could be a great investment. By simply tearing this small 1,500 sq. ft. 1-story house down and building a 5,000 sq. ft. 3-story home, this fact will make this house, or any other ocean, lake, or beach house a great investment.

Buying smaller houses and tearing them down and building bigger houses is a trend in beach front areas, which creates massive amounts of wealth especially renting them out, which creates generational wealth.

These rod-iron fences which were very affordably put up are taking on a different look now by using very affordable plywood inserts and then stained which gives it a very rich look creating a very high demand.

These fences are creating a lot of profit for welders and for people who have a trade in using hammers, nails, and screws or light-weight welding, which is a very lucrative career for someone who is willing to earn six to seven figures a year.

MONETIZING GENTRIFICATION IN LAS VEGAS

Homes with 2-story ceilings can be easily converted to extra bedrooms making a 2 bedroom into a 3-bedroom house. So instead of looking for a 3-bedroom house, you can look for a 2 bedroom with a 2 story or vaulted ceiling saving large amounts of money in the process.

With California's high cost of living, Las Vegas is benefitting greatly with a large amount of California residents moving there. With the Raiders coming, there are more hotels being built to accommodate people who are looking for homes in the meantime. So, Las Vegas' residential property is seeing large increases in equity.

MONETIZING GENTRIFIFCATION IN NORTH CAROLINA

The house pictured above is in Fayetteville, North Carolina, and fell victim to imminent domain. The homeowners lost their front yard due to a brand new freeway. New off ramp and stoplights (pictured on the left) have homeowners so upset that they're most likely ready to sell, which opens the door for opportunities seeing that this house will one day be torn down and replaced by a gas station or shopping center. The zoning is about to change, which opens the door for wealth building opportunities. Newly built freeways are the pathways to wealth building. I always say, "Follow the freeway!"

Premanufactured homes have become very popular because of rising housing costs. So you have a large amount of people buying land and dropping these premanufactured homes on them. Financing is much different from buying a traditional home because these can be removed from the land. Technology has made these so nice that sometimes you don't even noticed that these are premanufactured homes. These are very affordable no matter what state you live in.

An amazing sign that gentrification/new development is coming is when you start seeing places like this that rents affordable offices. Every construction site has a portable office on its worksite. So when places like this pop up, that tells me that major construction is coming this way because these places want to be near all of the action. Portable offices as well as equipment rental places and new Home Depot, Lowes, etc., popping up is a sign of gentrification and redevelopment.

A small house like this sitting on large parcels of land half an acre and up sitting on any major highway is a perfect location to invest in. These houses can be used as public storages, car lots and all types of similar businesses and have major exposure due to being on a highway with high traffic.

STOLEN IDEAS

So let's talk about some business ideas. Who comes up with these ideas? Is it large corporations, organizations, employers, or their employees? The answer is: employees! Employees like to make their lives easier. The employers don't care what you're doing, just get it done. The employees come up with better ideas that will make the job work more efficiently and save them some time to work things out for them very well. But here's the thing, most employees don't realize they've signed a waiver that if you sign up and work for a job, that job owns any idea that you come up with.

So, if you come up with a great idea for a new business that you thought about from the business that you work for, your boss will be the owner of that idea because they inspired you. So what's happening with that? A lot of people's ideas are being stolen. So keep your own ideas, bring them to me and let me help you start a business with them. Invest in yourself!

What's the most stolen thing on the planet? Of course, we just said it, an idea. We come up with an idea and other people steal them from us or give us peanuts for them. A billion dollars could be peanuts, and they go off and make billions and trillions of dollars. We should sell our own ideas ourselves.

So what ideas paid off huge returns? The Benihana owner came from Japan. He realized there was no fried rice in this country. He decided to open up a fried rice place in America; now he has thousands of them all around the world; this guy's a multibillionaire.

With The Bedford Group, some black people decided, "Hey, we want to build houses for our own people." Now all of a sudden guess what? Billion dollar black-owned companies are being created.

Apple computers decided "Hey, we want to start this thing called a computer, a personal computer." He later died a multi-trillionaire.

Facebook guy decided, "Hey we're going to make a social media app." Now he's a Billionaire.

Roscoe's, the chicken and waffles guy decided, "Hey, I'm going to make some chicken and put some waffles with it and sell it for breakfast." They thought that was stupid, but guess what? He's a multibillionaire.

Candy Crush video game, everyone thought that was a stupid game but people got addicted and fell in love with it and play it every day. Candy Crush sold for $5.9 billion dollars.

Somebody decided to make a video game about someone in the hood going throughout the neighborhood busting windows out of cars, walking his dog, graffiti on walls, and beating people up. They decided let's make a video game called Grand Theft Auto, which is now a billion-dollar franchise.

Come on you all, there is no such thing as a bad idea. There is no such thing as a dumb idea, somebody is going to pay for it. Because all of these things seem to be dumb ideas when they got started, now people all around the world are spending billions and trillions of dollars a year buying them.

THE BLUEPRINT SOLUTIONS

Inspire your Children:

a.   Attend Black Business classes online or in person taught by African American/Black instructors.

b.   Enroll in free Home Depot/ Lowe's classes.

c.   Start a business early like a Lemonade stand, gardener, shovel snow, trash can monitor for neighbors.

d.   Teach them to negotiate/ Barter.

e.   Show them how gas prices are affordable and expensive, good or bad gas stations, and why they are that way.

f.   Share your bills with your children let them know how much things cost.

g.   Teach them the proper way to use social media and create a following early so that they may charge others to post on their page, and they can charge businesses to monitor their pages.

h.   Have them intern at a local business.

i.   Always speak about what things cost especially while shopping for food, clothes, cars etc. Ask them for their opinions.

j.   Take your children to open houses from regular priced homes to million dollar neighborhoods. They should know why things are so cheap or why they are so expensive.

k.   Show them construction sites and monitor the progress on them.

THE BLUEPRINT SOLUTIONS
continued...

If your children are going to attend college, only allow them to take classes that they can start a business with like these to name a few:

- Attorney
- Dentist
- Film maker
- Scientist
- Architect
- Civil engineer
- Contractor
- Mechanic
- Electrician
- Welder
- Painter
- Real estate agent, broker
- Airplane pilot (private charter company)

If you currently have a job, you should already have enough information to start some type of successful business like the following:

- Security guard = security company
- Janitor = maintenance company
- School teacher = start a small private school
- Mail person = shipping business
- Limo driver = limousine company
- Bus driver = party bus owner
- Auto mechanic = auto shop owner
- Plumber = plumbing business

The best teacher on the planet is a person that's running a business that teaches people how to run businesses. So as they age, they can keep their businesses going. When people leave houses to the churches, the churches make bigger churches. And people who leave their money to business owners, the black business owners can start hiring more people within the community and thereby create jobs and help build stronger family structures.

In business I realized at a young age that it's better to be married as a business owner and your spouse can help you build your business(es) together. So if I were in college, I would probably be dating someone in law or accounting or someone who can help me build my business so I won't have to spend extra money with contractors, etc., My spouse can handle that work. That's a perfect place to find your spouse because we need people to help us grow our community. This is the purpose of this book, to secure our communities and build stronger bonds for future generations to invest in.

I suggest people look at successful businesses like Simply Wholesome in Los Angeles and Tassilli's in Atlanta to invest in, even myself. I travel the county educating people. I find it difficult to find places to speak. So I would like to open up places for speakers around the world to speak and they will need housing to stay etc., but I would open up facilities in every major city to have a place for entrepreneurial trainings.

I would also be able to teach the trades: plumbing, welding, automotive, etc., that I'm currently doing on a smaller scale, which I would employ at least 20 people around the country that would create jobs and I would train them to do like I did, start businesses and partner with your neighbors and friends and watch the community grow.

WEALTH THROUGH THE TRADES

So let's discuss affordable places to live. Buy land and build your own house. You can build for less than what it takes to buy a house. You can move into the suburbs and get a better deal than you can in the city area, or inner cities. Or you can own your land and put an RV on it and do very well. Live cheap until you build a house, or you can move into an RV park, buy you an RV and move into an RV park that's very affordable. But the most affordable place to live is on a yacht in a marina. So when I tell people about the art of negotiation just like I told you about the $2,000,000 mansion for sale, you can find a yacht that was sold for $1,000,000, and now the guy wants to sell it. It's been on the market for over a year and he wants to buy a yacht, for example.

You don't have to buy a huge yacht, you can buy a smaller yacht, a regular yacht, a regular boat. A lot of people are moving on the water because it's very affordable. If you have a 30-foot boat you can put that on the water in California for $210. That's $7.00 a square foot which equals $210 a month and that's with utilities paid. So a lot of people don't realize that there are a lot people living in their boats.

We just see them and we think, "Oh you live behind these nice restaurants you're doing well, you're the man, or you're the woman, or you're the big dog." But we don't realize is that some of these people are struggling but they're smart. They're saving money by living on their boat in the marina. We think people who live in RV parks are poor but when you go look at some of these RV's you'd be amazed.

This is what the new "trailer trash" looks like in 2018. So, what's the new wealthy look like right now? I tell people we have to start putting our children in the free Home Depot classes because it's not about being a doctor or a lawyer anymore, for the majority. Nothing's wrong with being those professionals because we need doctors and lawyers, but we put our children in school to learn a bunch of stuff that doesn't make the wealthy; but guess what? The people that have trade skills, you don't find these classes in colleges, you'll find these in trade schools. You don't find these in 4-year colleges, you find these in trade schools and regular 2-year colleges. But most of the time they pull all these classes out of the school system.

You used to learn all these careers in junior high school: carpenters, masonry, welders, plumbing, heat and air, HVAC, electricians, gardeners, or owning a construction company. These are the people that are going to become the new multi-billionaires. Why? Because we're building, we're in a construction age. Donald Trump is currently the President and we're redeveloping America. So I tell people construction is where it's at in building houses. Masonry, a lot of people think, "Oh, I can't make any money learning how to stack bricks." You can learn how to stack bricks in an hour. Just do a couple of little walls and next thing you know, you know how to lay brick. Go to Home Depot and ask them how to lay brick. Go there and take a free class on anything that has to do with construction. Go to there and take all of those free classes and then you start your own business doing this. I'm a celebrity car builder, I built cars for celebrities. Can you imagine if I built homes for celebrities or brick walls for celebrities?

Let's look at this brick wall here. When you say this little brick wall in a small neighborhood may be around $10,000 and $15,000, what will happen if you go to a celebrity and build their brick wall? The owner would say they need a brick wall around their big mansion and they may quote him $1,000,000 and materials only cost you $50,000. Carpenters are making a lot of money. Masonry people who put brick down make a lot of money, too.

LEARN HOW TO CREATE A DRIVEWAY...

So, those same people that put that driveway in front of a house in the hood at $10,000, are the same people who could build a driveway in front of a mansion. For example, Jay Z and Beyoncé's new $88,000,000 mansion has the same material and same knowledge by someone who could lay this down, it's just a little more detailed but charged a lot more money. Jay Z and Beyoncé made history by purchasing that house. So think about it, if I'm a celebrity car builder, I built cars for celebrities, why can't we have celebrity home builders? You know black people who build homes for celebrities, just like I did for the cars builders, make a lot of money. That's why it's important to learn a trade.

LEARN HOW TO BUILD A FENCE...

The person who builds a gate in Compton, California, in the hood in the black community could be worth about $10,000. You take that same gate you put it in front of Jay Z and Beyoncé's $88,000,000 mansion or whatever they paid for it, that gate will be worth $1,000,000. Same material, approximately the same amount of time and same look. Carpenters make a lot of money. Welders make a lot of money, too.

With these cabinets here, you can make these at home in your garage, for that same $88,000,000 house and it can cost you an easy $4,000,000. You can make all of these cabinets sitting at home in your garage, from a YouTube video or a Home Depot class.

The same rules apply when it comes to these doors. You can make these doors while sitting at home in your garage. That's why I tell people they can become a celebrity door builder, a celebrity home builder, or a celebrity plumber.

The same people who put this plumbing around this garden are the same people who can put this plumbing in this million-dollar mansion. How about you become a plumber to the stars?

So when we take our young people on field trips, we have to start taking them places where they can use real-world information. Our young people are coming home with college degrees and can't fix a door handle, can't work on a car, or repair their houses. This is why our young people who are generation X or Z are starting to get the homes from the baby boomers or their grandparents and they're selling them. They don't know how to do the work and over the years end up letting the houses go because they didn't know how to fix them. They didn't have the money to pay to fix them. So they decide to sell the houses. If everybody had a basic class in construction, we could keep our own houses up to date and most importantly we can make a lot of money off of them. These carpenters are making all of the money because they are not having to move away. If they move away because of a job, it's because someone is paying them more money, and it's definitely worth moving.

I took these young people on a field trip to go look at this brand new house being built and these young people called me every day telling me the new status of the new house that was being built. The roof was being put on one day, they emptied the port-a-potty, and then they dumped the bin. They told me that they knew so much about it and all the time we spent in front of this house was approximately 20 minutes. I just explained to them what the name of everything was and what things cost and then these young people knew everything about it.

There are so many business opportunities in construction as well as business ownership. I tell people all of the time, if you are a carpenter and you learn how to build a house or if you learn how to build a car, or work on a car, and you come up with an idea to make a nail, hypothetically speaking, you can create a better design for making a nail that is easier to hammer. Next thing you know, you start selling that nail on the market. That nail may cost about you $6.00 to manufacture.

So you start to make that nail and all of a sudden you're making nails and you're probably paying a penny per nail, and you're selling them for $.25-$.50 per nail. Now you've revolutionized the nail industry with this brand new design of a nail. Someone offers you a billion dollars for your nail and you say, "No, I'm going to keep it." Then you decided you're going to make a hammer to go with that nail; you're going to make a tool belt to hold that hammer and those nails; and next thing you know you decide to make your whole tool company; and then you're hiring everybody in your community, your church, everybody you know, giving them jobs with people going all around the world selling your nails. This is why it's important to become manufacturers. This is why it is important to learn a trade. Trade in carpentry, trade in mechanics, or any trade because you come up with ideas.

THINGS YOU SHOULD KNOW

A trade is a pathway to wealth and wealth can make you a billionaire manufacturing things. But you can't become a manufacturer, if you don't have some type of a trade.

So these are some of the ways money is being made now. We are moving away from the traditional education. People are going to school to get their Bachelor's Degrees and higher education, but these people aren't getting hired. They're using crazy language saying they have to compete. They have to do this, they have to do that; they have to work hard, they have to go back to school and get another class to learn something else. Well, guess what? People building houses didn't go to traditional schools.

They get on-the-job training, and they are becoming multi-millionaires, if not billionaires by designing nails and hammers and screw guns and all of these nice cool devices. Again, you can't get any of these ideas if you don't have the experience or if you don't have that type of trade to go in there and do these things. And remember if you work for someone, your idea is their idea. But in this industry, if you become a carpenter or a mechanic you start your own business and your ideas are your own ideas, you become a billionaire designing nails.

A woodshop teacher in Compton, California, reached out to me after hearing me talk about going back to the trades and building tiny houses. After going in and seeing that they were building tiny houses, it gave me hope knowing that this school was headed in the right direction. Now after getting there and discussing things with the teacher, I discovered that he was doing this out of his pockets. They weren't really trying to fund a tiny home project in the school district even though they know there's a building boom going on in the city of Compton and California. The children would be guaranteed jobs with this trade if they learned this trade. Plumbing, dry wall, roofing, electrical, hanging doors and windows, painting, carpentry, masonry, etc., but most importantly I knew they'd grow up to become billionaires if they started building these homes with their classmates seeing that the median home prices are starting off at $600,000. I gave a couple of presentations to the children in the classroom inspiring them to learn this trade so they would never have to work for anyone if they started a business with these skills.

WEALTH IS A CHOICE

Every year millions of people around the world decide to put their children in college basically saying they want their child to be an employee which equates to them earning from $40,000 to $60,000 annually. A college student will go to school for four-five years to earn a degree of some sort. Some go even longer to learn something that they can download off the computer for free. On an average let's stick with four-five years. After graduating, students will spend at least another year looking for a job. After finding that job they're going to spend another year as an intern, which means college graduates will spend on an average of eight years to learn how to get a job earning $40,000 to $60,000 annually.

What parents should be saying is that they want their children to be wealthy and start some type of business and invest in real estate. It is a proven fact that anyone with a product or a service who dedicates that kind of time for up to eight years into anything will become successful earning them on an average of six to seven figures a year.

Let's use a book for example, if a person writes a self-published book about something that they've experienced in their life, they spent a total of two to three hours a day finding out everything about writing a book from manufacturing it themselves to distributing it themselves finding out about where the Black book stores and businesses are around the world. And creating a relationship with them asking them can they use their facility to have a book signing. They spend maybe four hours a day learning social media keeping up with all the new social media and trends to help with their advertising. Using the same social media to find out who owns black bookstores and businesses. Also finding regular everyday people who are social media famous like local rappers, comedians, and musicians that they grew up with that have 75,000+ followers especially beautiful women who like to model. All of these people can have upwards of 1,000,000 followers because they've been on social media for so long promoting themselves. So simply by asking them to like your book and post a picture of it on their social media to help drive more people to your page can help you earn thousands of dollars a year connecting with popular people on social media that your average person has no clue of who they are; they're just social media famous.

Most people start off with the book then it turns into an eBook and/or audiobook then they're doing public speaking branding themselves as an expert. So they travel the country speaking everywhere promoting their book. Then they start training others to write books themselves, new social media. Then they start charging people to manufacture their books for them because they are the experts, they know the game. Up and coming people will always gravitate towards them for coaching opening up so many different forms of revenue. It is a known fact after that kind of dedication and research you should be earning at least six-to-seven figures a year. Proving that wealth is a choice!

QUESTIONS

- What are your opportunities for creating wealth now?

- What does it take to become social media famous?

MANY FACES OF GENTRIFICATION

When listing property, sellers heavily rely on the selling agent to list their house and get it sold. The agent is the one who is responsible for advising the seller of what the property is worth. Because of gentrification some agents are pushing the market up by misleading the seller and telling them the house is worth more than market value making sellers happy with more money but they're never truly informed on what the house should be listed for. So, if the house should be listed at $500,000, the real estate agent may recommend to list it at $625,000. Now one thing that happens is the property may not get any real offers on it because other agents see that the properties are overpriced so they won't waste time by putting an offer on it. After sitting on the market for a while the seller starts to get discouraged and reaches out to the agent about solutions on selling it fast. Now the agent has a buyer in place to buy the property at a reduced price. The seller gets excited and decides to take the $450,000 or less.

QUESTIONS

- Are there any signs of gentrification going on within your community?

- If so, what are the signs?

RANCH PROPERTY BOOM IS GOING GREEN

Because of what's going on in the world with all the GMO foods and the high cost of living, more people are turning green, which means they're deciding to build sustainable homes off the grid. People want to plant gardens in their backyards. They also want to run their businesses from their homes. While running my car business I spent thousands of dollars a month to store cars all over the city which was a huge inconvenience to myself. Then someone explained to me that I should purchase a piece of land and build my own home on it and consolidate my business into one location. I found 2.3 acres with the small 2,100 square foot house on it with a large building in the backyard that I could run my business out of. I was able to put my tow trucks, my car carriers, my trailers, and all of my junk yard cars in the backyard without even bothering my standard of living; some people didn't even know all of that was back there. It saved me thousands of dollars a month by reducing all of my bills down to a low cost of a mortgage.

The SQUARE FOOTAGE GAME

In order to really monetize gentrification you have to know the square footage game. Meaning you need to know the difference between an 800 square foot house from a 6,000 square foot house. Most people who are looking for a very large home may say I want a 7-bedroom house. Well the problem with that is you can have seven bedrooms in a thousand square foot house. They should be asking about a 6,000 square foot house instead. Others looking for a smaller home will ask for a two-bedroom home. Well you can put two bedrooms in a 600 square foot home. I've been in homes 4,000 square feet that had two bedrooms. When interested in purchasing a home, always ask them, "What is the square footage?" to get a clear *overstanding* of the square footage. I'm not interested in looking at anything less than 3,500 square feet. What about four bedrooms? I also know that I'm not interested in anything over 6,000 square feet at this time. If it is 6,000 square feet, it should have an elevator inside because I'm very aware of how the square footage works. When dealing with large homes of 4,000 square feet and above that means I should have more than one air conditioner and heater. It should have four air condition units and four heating units. If I'm home alone in my office and would like to have the air conditioner on, I don't want to cool the whole house; that would be a waste of money. So having different units can keep the bills down.

QUESTION

- Why is it important to ask about the square footage of a building?

FUTURE VALUES

When buying a house I'm not necessarily looking for something that's appealing to my heart. I'm looking for something that will have great future value; meaning, a small house on a large 8,000+ square foot lot close to a mall or something popular. Over the years, buying trends have changed. Back in the early 1900s, families had two bedrooms and one bathroom houses 800 to 1,200 square feet with one-car garages.

Now, families are looking for much larger houses like a 7-bedroom, 6,000 square foot home with a four-car garage. Now the big secret is most people don't realize that back in the early 1900s when they were building those houses, most houses came with small lots 3,500 to 5,000 square feet. They have pockets of homes considered small ranches with large lots 7,000 square feet and up. Those are the most valuable lots. So I'm only looking for houses with 8,000 square feet and higher lots. In the future I will tear down a small 1,200 square foot home and build 6,000 – 7,000 square feet homes or apartment buildings. That's the future value I'm looking for.

QUESTION

- What trends have you noticed within your community?

FORCED SALES

There are many faces of gentrification designed to force people to sell their properties. A major one was cities coming out with major strategies on how to force people to sell their properties indirectly. Those cities decided to outsource a department designed to go after investors that own multiple apartment buildings and houses. They figured out which one the owner lives in and then send all of the tenants renting from them a letter called REAP, which means Renters' Escrow Assistance Program. Cities are using this tactic all around the country but probably under different names. What they did is request the renters to send them half of your money and tell the owner to contact the city. The owner would then contact the city, asking why they're taking the owner's money and the city would say, "Oh, we meant to call you. So we would like you to meet us at property Number 1."

They'd go in from the checklist to find a lot of petty things that needs to be fixed like replace an old screen, the hood vent is too old, you need a new ceiling fan, etc., but that would easily run up $4,000 and up in repairs on each property and if you didn't make the unnecessary repairs, they will put a lien against the property. Now multiply that times three other properties that could run an easy $20,000 and up in repairs. Now a lot of people may decide that they're tired of dealing with the city and are not going to make the repairs. A lot of people may decide to put the properties on the market for sale because they don't have time to play with the city. Just because the owner sold the property does not mean to REAP goes away. The REAP stays with the building so that an up-and-coming buyer may not be aware of the REAP status and find themselves in the same situation. And it's not that easy to take REAP off of a property. The city is using that as a way to control people's profits and put money in their own pockets. The goal is for people to tear down the old properties and build brand new ones; that's how you get rid of the city. Some would-be investors are not aware of that and will keep dumping money into these old properties.

The city's main goal is to tear down all of the old properties, houses, apartment buildings, commercial buildings, etc., and have new large scale properties built. The reason why is because it is predicted by 2050 Los Angeles county will have 10,000,000 more people. So they have to make room for them. So you have a lot of apartment buildings being dumped on the market for sale. You should be very careful when you see things like that. If you see a lot of new construction around the property that is for sale, just know that it's a strong possibility that the city may harass you. Goal being to get you to tear that down and build something bigger.

RECOGNIZING AN OPPORTUNITY

Years ago I stumbled across a new area where the property values were down. I found some two-story townhomes that were going for about $50,000 to $70,000. After my deposit, my mortgage payments were going to be about $300 a month. Now those buildings didn't look that great; in fact, a close friend strongly encouraged me not to buy those because they weren't easy on the eye but because I have the eye of an investor I recognized an opportunity. What I realized was the market rent in the area for a two bedroom, two bathroom townhomes were $400 a month. Keep in mind they have four units in there so that's a total of $1,600 per month and my mortgage was only $300 per month, which gave me a total of $1,300 per month after bills to walk away with at least $1,000 per month in profit. Some people get caught up in the emotions. I wasn't purchasing the property for me to live in. I was purchasing the property as an investment for someone else to live in. That friend said that they would never live there so why would anybody else? That's not the case. Never let your emotions get in the way of an investment. Those properties went on to cost over $600,000 each over the years.

QUESTIONS

- Are there any investment opportunities within your community?

- Do you have an eye of an investor?

APPRAISAL

Another way that gentrification becomes very relevant is when an appraiser goes into an African American community, low ball the property values by telling the homeowner that your property value has come in low because of the drive-by shootings that were on the corner, etc. So the homeowner becomes desperate and says, "Yes, that is the reason why I'm selling so I am going to take a hit and lower the cost of my house." But the reality is that homeowners should make sure that the appraiser knows about all of the new corporations moving in the community like in Los Angeles. We have a lot of technology companies that are moving five minutes away like Google, Space X, Yahoo, etc., are all new residents. So many billion-dollar corporations are located within five minutes from your house. Not knowing your area can greatly work against you. Appraisals are being done so you make sure you know as well as the appraiser that your home is in the rapidly growing area due to all the multibillion-dollar corporations moving in.

QUESTIONS

- Why is it important to ask your appraiser questions?

- What big companies are located in your community?

LEAVING A WILL IS IMPORTANT

Most of our baby boomers purchased homes in the late 60s and early 70s for $12,000 to $25,000. Now those same homes are worth anywhere from $600,000 to $1.2 million depending upon the location. For example, homes in Watts, California, had no value because of the crime in the area. But now because of gentrification, purchasing the Jordan Down Projects' value in those neighborhoods have skyrocketed. The same thing is going on all around the country. Undesirable neighborhood values are skyrocketing overnight. So, now you have family members who did not want the real estate their parents were leaving them in the bad areas that you once grew up in. One of your family members decides to take it. Now you have the other family re-deciding to come back and fight for equal ownership of the properties that they at one time did not want at all. Big corporations moving in helped them to change their minds. This is one of the reasons to have a will that has been carefully drafted out and also put in writing and recorded on video.

QUESTIONS

- Why is it important to leave a will?

- Does your family have a will?

PERCENT GAME

I want to talk about the percent game. Every house that's being sold pays a 6% split: 3% to the seller's agent and 3% to the buyer's agent. The 6% split is normally known as a commission paid by the seller of the home. Now what the sellers don't normally realize is when they put their house up for sale saying that they're going to pay a 6% commission on that house; it gives 3% to the seller's real estate agent and 3% to the buyer's real estate agent.

Also, what the sellers don't know is the agent plays games during gentrification that you will never know about. The real estate agent representing the seller without the seller's knowledge will put the house on the market, paying a 2% commission instead of a traditional 3% commission. Now most agents will never accept a 2% split because they know they're supposed to get a 3% commission. So other real estate agents will never show that property. If the customer finds out about the house, the agent will lie and tell them something that will discourage them from buying it. All based on the fact that they're not going to make a 3% commission on that property. So that becomes very discouraging for the seller. They're wondering why their house is still on the market after all of this time. Then the seller will go back to the real estate agent asking them to lower the price so that they may get a fast sale.

When agents do this, their goal is to stress out the seller in an effort to reduce the price of the home so that they can bring in their investors to buy the home cheaper than market value. This is why people should know that some real estate agents are not their friends. This usually happens in prime real estate areas that are being gentrified. There's a feature on the MLS (Multiple Listing Service) exclusive to real estate agents only that will only show pertinent information like the commission split, the access codes to the door locks, etc., and only real estate agents have access. In that system, there's a tab where an agent can show that they only want to look at houses paying a 3% and up commission only. So, if a homeowner's house is listed below 3%, most agents will never see it. Which means the potential buyers will not know about the house unless they see a *For Sale* sign in front of the house.

QUESTIONS

- What is the average commission split?

- What does MLS stand for?

DURING FORECLOSURE

During foreclosures people try to sell their houses early before losing the house and totally mess up their credit by doing a short sale. So investors and real estate agents watch for houses that are falling behind on their taxes or are in pre-foreclosure. So when they put a house up for sale they come in and lowball those desperate to sell. So, an easy way to deal with that is to have your listing agent change the pay to 4% to whomever brings the new buyer. You pay your agent 2% unless they double ends the deal. By paying 4% to the agent who finds the new buyer will bring in a large turnout of potential buyers at regular price. Reason being the average commission paid to each side is 3% each. So knowing that you're going to make 4% equals more money to the selling agent. So, of course, agents are going to pay special attention to your property at all costs. They're going to show your house first and strongly encourage their potential buyers to purchase your home at full cost. Houses that are in foreclosure are usually lost to the banks because you only have a certain amount of time to sell. Most agents don't want to deal with foreclosures because there's no incentive on selling one. That's why you pay 4%. That's how some people avoid losing their houses to the bank.

EARLY LESSONS LEARNED

My friends and I used to take our boats out to Lake Havasu, AZ, every holiday and stay in rental properties which became very expensive. So I wanted to start buying houses so we can store boats and then one day we got smart and decided we want to build some condos on the beach that way we can keep our boats on the water at our condo. Before we got started, we realized someone else had the same bright idea and bought some prime real estate right on the lakefront. We decided to buy one from them so we pulled our money together knowing that the median home price in the area is about $400,000. So we were quite sure that the lakefront condos would run about $250,000. We put money aside to buy one. When the condos were finally finished we were very excited to put our bid in to buy one with cash but before we could, we found out that the owner had his friend buy the first three for $600,000 cash each. That was a major power move saying that those condos would never appraise for nothing over $250,000. So by paying cash that immediately raised the property values on the rest of them that sold for about $700,000 and up. Before they were all done they were selling at almost $1 million dollars each. Needless to say we never bought one and decided to go back with our original plan of building. A valuable lesson was learned; when you are a builder and you want to raise property values, all you need to do is buy the first three with cash and that will set the comps for the rest of the area. I see people using this strategy all over the country.

QUESTIONS

- Would you consider investing in land by water?

- Why should an investor purchase three properties at one time?

BEING SET UP FOR FAILURE

Your landlord is setting you up for failure by renting to a person long term. The longer you rent the higher the property values continue to climb making it more difficult for a person to buy a home or a commercial building to run their business from. Everyday millions of people search for a home to live in or a building to run their businesses from. Most have a choice to make because the properties are for sale or lease. If you buy it, for example, it may cost you $1,000 per month with a $6,000 total down payment and the value of the property may be $160,000.

Now on the other hand if you rent, it may cost $1,800 per month or $3,600 and up to move in depending upon your credit and every year the rent will go up 2%. Due to the major disruption and all of the power moves going on in the world especially gentrification that $160,000 property may climb up to $1,000,000 in the next five years. Now an owner will become rich if they choose to sell it but can create generational wealth if they choose to keep it by passing the property down to their children allowing them to continue to run the family business or just sit back and collect rent for the rest of their lives.

While the person who chooses to lease or rent the property will only have sad thoughts thinking they should have purchased five years ago instead of renting are not able to create generational wealth by leaving their property to their children. But will most likely be forced to move due to rent increases.

"It's better to buy a shack than to rent a castle; only the shack can be left to your children." TJ

THE LAND GAME

All over the country land has been increasing in value over night purely based on speculations. Real estate agents are working together to raise up the price of land all over the country just to make more money. Most land is not sold through a real estate agent because the values are too low. Banks won't finance a piece of land less than $100,000, so land prices have historically been low where a person can pay cash for them. Because of this most realtors will not take a land listing causing the owner to list the property to sale by owner (FSBO which means For Sale by Owner).

Some real estate agents have devised a plan to raise the land values around the country. When I see land listings that's not sold for over a year, I immediately gravitate towards those properties. Meaning a listing that's been on the market for over a year means no one's interested in purchasing that land because there is a lot of value out there. That opens the door for me to negotiate. So now some real estate agents started working together convincing sellers that their $1,500 piece of land is really worth $20,000. I noticed this when a lot of properties were listed straight across the board from various real estate offices for $20,000 and up. I thought that was kind of odd until I dug deeper and started contacting some of the agents asking them what exactly justifies a price so high.

So the agent preceded to tell me in California City, California, they're having a real estate boom. They also talked about the medical marijuana being legalized in Kern County which is also making the land prices go up. So normally real estate agents are not used to being questioned, they're used to dealing with people who take their word and says let's go for it; sell me some land at $20,000. So I proceeded to ask the realtor where exactly is the boom and what's causing it and where are they building them because I just drove around the city and can't find any new construction to justify your boom and high prices. Where exactly are the marijuana growers building their facilities so I can buy some land near there as well; of course, the agent couldn't tell me anything. She didn't know, which meant it wasn't true. They were basically selling land on rumors to drive the market up. So, I asked her why is she stating something that she couldn't verify. And she proceeded to struggle. She said that I wasn't serious. I said I'm very serious when you can show me where these medical marijuana companies are building their facilities or where Mike Tyson is building his large facility or show me on title where he purchased the property, I'll buy all the land around it that you can find me. Of course she couldn't so that showed me that some real estate agents all over the country are selling land based on speculation just to drive the prices up. Those who don't know or understand real estate will be bamboozled by paying more money. What they should be doing is asking the agent what was the date and price of the last three properties sold. So you may find that the last three properties sold over a year ago substantially less than what they're asking for, and I would base my offer on the last ones sold and not the speculation price of new corporations moving in. This shows the reason why we need more trustworthy black real estate agents. Every family should have its own real estate agent. Because there are a lot of people who cannot afford to pay $20,000 for a piece of land and look forward to purchasing one for $800 or an affordable house. Because of the housing crisis, real estate agents are realizing that more people are buying land and starting to build their own affordable homes.

"Always remember it's cheaper to build a home than it is to buy a preexisting home." TJ

In an effort to continue to make money, I see real estate agents focusing more on bringing the land owners to sell their land higher, which will drive the land prices up around the country. I have the eye of an investor, therefore, I realize the higher the home costs that African Americans are being priced out of real estate ownership. So, I've been traveling around the country telling and training our communities how to go back to our roots of building. This is the only way of maintaining home and business ownership.

"We are moving into a construction era where the rich will buy and the poor will be forced to build." TJ

We should have never moved away from our roots of building. Just like during Black Wall Street, they would say that whites would burn down the black houses and the blacks would come together as a community and rebuild them brand new using our own hands and knowledge that has been passed down from generations.

EDUCATIONAL LIMITATIONS

When someone says they put their child in school to get a good job, I realize that you're saying that you're going to limit yourself to making $60,000 - $100,000, hopefully. You say you're going to put your daughter in school to become a doctor. Well, your daughter is only going to make $160,000 to $300,000 for five or six years or seven years of school. Now if she decides she wants to become a specialist, such as a brain surgeon or a skin surgeon, then now we are talking about how she can become a millionaire, or he can become a millionaire, but the majority of them are not. So again most people who become doctors are making somewhere around $160,000 to $300,000. That's barely enough to live in California, New York, and similar places. I tell people all of the time that if you want to become successful, you have to ask yourself do you want to become successful or do want a good job? If you want a good job, you're limiting yourself on what you can make and you're going to have to move out of state, most likely.

But if you really want to become successful, you need to hire a coach. If you want to become a millionaire or a billionaire you need to hire a coach, someone who's been there and done that. And that is exactly what I do, I've been there and I've done that. I offer classes on entrepreneurship, automotive, classic and custom cars, low riding manufacturing, land development, and also the hair industry. I have a youth mentorship going on, too. If you decide you're not going to go to college, you need me to help you and I can provide the coaching you need to become successful.

I can offer you coaching services whether it's for 30 days or it's a year. I'll sit down with you and work with you and give you a price. I have a lot of people coming in now that are saying they're not going to put their child in college; they're going to go ahead and pay me to teach them how to become successful millionaires. You will see why I'm in so many magazines. You will see why they've decided that they want their son or daughter or themselves to become millionaires. So they're going to let me coach them and train them and guide them through all the way from real estate, to business ownership, to what to do with money, and most importantly how to keep the money. So if you decide that you would like to take one of my training courses, again you can reach out to me. No more becoming regular people, no more trying to get a good job because in America a good job is no longer enough. You will not be a homeowner with a good job in America. You have to be a successful business owner. Also, when you have some time, go on to my social media site on Facebook - Thomas TJ Loftin and click "follow," and "like." I have a business, go to my public figure page and click around and follow me. That would be great. Go to my Instagram page and click "like" and the same with Twitter. Go to my website and keep up with me and see where I am in the world at www.ThomasTJLoftin.com.

There is a calendar on there. Also look for my *TJ's Wealth Academy.* I have an academy where you can go on my website and download all of my webinars and take my training classes because I have created these training classes to teach people about creating wealth. So it's business ownership from the perspective of a business guru, a guy who has been in the business for decades that's helped a lot of people to become successful that coaches celebrities, and more. So I've finally opened up to the public to help whoever wants to become successful. Let's sit down and talk about it and let's make this happen because right now we can't afford not to. Now you're learning how things are changing. It's no longer about getting a good education and getting a good job because both of those are becoming behind us. We've moved beyond these things. So here's my information: 310-619-3954. Reach out and give me a call.

I also have a "Go Fund Me" page, too. Rich people tell me I need to push that page because in my spare time, when I'm available, I love to go to the schools and talk to the young people. Go to my YouTube page at Thomas "TJ" Loftin and click subscribe and go to all my social media and click subscribe. And if you can, donate to my Go Fund Me account because I love to go to schools in my spare time. Here's the thing, K-12 schools don't have the money to pay to get a person like me to come in. So I volunteer to go to those schools out of my own pocket. So whenever I go to another state, I always reach out to people and ask them what schools I can go to when I want to speak to the young people and talk about entrepreneurship.

If you want me to come out to you and speak to your city in your state, I can do that, too. Reach out to me and let's talk about the fee that I'll charge you to come out there, but reach out to me and let me know when you're ready to create wealth or help someone help your community to create wealth.

I've helped a lot of people become successful. A lot of my past employees are millionaires now. A lot of people are poster millionaires if not more successful than that. So now if you'd like my help I can help you to become successful, and you don't' have to think. A lot of times people think that they need things that they don't need to become successful. A lot of people think that they don't have enough money to become successful. A lot of people think they don't have enough money to buy a house. The whole thing is most people have the wrong information. So again I offer coaching that will completely change your whole way of thinking, your whole perspective, and teach you which way to go, and then I also have coaching for people who prefer my college package. People who want to learn how to create wealth and want me to hold their hand. So it's not like the coaching with other people, it's a different kind of coaching where I sit down with you 1-on-1 coaching or we can talk over the phone. We can do Facetime no matter where you are around the country and we can talk a couple times a week, 3, 4, 5 times a week depending on which package you take and make sure you are successful.

My coaching services range from anything from real estate to the auto industry, to straight up business. I've helped my family members open up businesses, very successful businesses in the hair industry and so many other industries. I'm not just limited to real estate and cars I am a very universal business person. If you have a bright idea that you came up with, you could talk to me about it and I can help you with it. I've got a lot of things coming up. We need more manufacturers because that's how wealth is created in our communities. If we have more manufacturers then we'll have more jobs, and we'll have more billionaires because manufacturing is a pathway to wealth.

"Having a trade is a pathway to manufacturing and being a manufacturer is a pathway to wealth." TJ

CPSIA information can be obtained
at www.ICGtesting.com
Printed in the USA
BVHW011923130719
553387BV00004B/8/P